Lecture Notes in Mathematics

Edited by A. Dold and B. Eckmann

1038

Francis Borceux
Gilberte Van den Bossche

Algebra in a Localic Topos with Applications to Ring Theory

Springer-Verlag
Berlin Heidelberg New York Tokyo 1983

Authors

Francis Borceux
Gilberte Van den Bossche
Département de Mathématique, Université de Louvain
2, chemin du Cyclotron, 1348 Louvain-la-Neuve, Belgium

AMS Subject Classifications (1980): 18F20, 18C10, 16A64, 16A90

ISBN 3-540-12711-9 Springer-Verlag Berlin Heidelberg New York Tokyo
ISBN 0-387-12711-9 Springer-Verlag New York Heidelberg Berlin Tokyo

Printing and binding: Beltz Offsetdruck, Hemsbach/Bergstr.
2146/3140-543210

0. INTRODUCTION

Sheaves of algebras on a topological space appear in many problems in mathematics and their interest has no longer to be demonstrated. The purpose of this publication is to study the localizations of the category of sheaves of $\mathbb{T}$-algebras, where $\mathbb{T}$ is a finitary algebraic theory, and the extent to which it characterizes the topological base space. The techniques developed to solve these problems, applied to the case of modules on a ring R, provide new results on pure ideals and the representation of rings. As a matter of fact we develop our study in the more general and more natural context of a theory $\mathbb{T}$ internally defined with respect to a topos of sheaves on a frame (i.e. a complete Heyting algebra; for example the algebra of open subsets of a topological space).

We shall normally use the letter $\mathbb{H}$ to denote a frame and, unless stated otherwise, $\mathbb{T}$ will denote a finitary algebraic theory in the topos of sheaves on $\mathbb{H}$. In chapter 1, we recall some basic properties of the categories $Pr(\mathbb{H},\mathbb{T})$ and $Sh(\mathbb{H},\mathbb{T})$ of presheaves and sheaves of $\mathbb{T}$-algebras on $\mathbb{H}$ (limits, colimits, generators, associated sheaf, and so on ...). Reference is made largely to classical texts.

In chapter 2, we first study the Heyting subobjects of a fixed object in $Sh(\mathbb{H},\mathbb{T})$: these are the subobjects wich satisfy properties analogous to the properties of any subobject in a topos. This allows us to describe the formal initial segments of $Sh(\mathbb{H},\mathbb{T})$. If $u{\downarrow}$ is any initial segment of $\mathbb{H}$ and $\mathbb{T}_{u\downarrow}$ the restriction of $\mathbb{T}$ to $u{\downarrow}$, $Sh(u{\downarrow},\mathbb{T}_{u\downarrow})$ is a subcategory of $Sh(\mathbb{H},\mathbb{T})$ satisfying very special properties. We then define "formal initial segments" to be subcategories of $Sh(\mathbb{H},\mathbb{T})$ satisfying analogous properties. The Heyting subobjects of a fixed algebraic sheaf constitute a frame and the same holds for the formal initial segment of $Sh(\mathbb{H},\mathbb{T})$.

Chapter 3 applies the results developed in chapter 2 to classify the localizations of $Sh(\mathbb{H},\mathbb{T})$ when the theory $\mathbb{T}$ is commutative. We build an object $\Omega_{\mathbb{T}}$ in a topos $\mathcal{E}(\mathbb{H},\mathbb{T})$; the localizations of $Sh(\mathbb{H},\mathbb{T})$ are exactly classified by the Lawvere - Tierney topologies $j : \Omega_{\mathbb{T}} \to \Omega_{\mathbb{T}}$. A characterization in terms of generalized Gabriel - Grothendieck topologies is also given. Examples are produced. A counterexample is given for the case of a non-commutative theory.

When $\mathbb{T}$ is the theory of sets, $\mathbb{H}$ can be easily recovered from the topos $Sh(\mathbb{H},\mathbb{T})$: it is the frame of subobjects of terminal object 1. In chapter 4, we introduce a large class of theories to be called integral : sets, monoids, groups, rings, modules on an integral domain, boolean algebras, ... are examples of them. When $\mathbb{T}$ is integral, the frame H can be recovered from the category Sh(H, $\mathbb{T}$) : it is the frame of formal initial segments defined in chapter 2.

In chapter 5, we expound the results on formal initial segments for a classical finitary algebraic theory $\mathbb{T}$. The category of $\mathbb{T}$-algebras is simply the category of sheaves of $\mathbb{T}$-algebras on the singleton. The frame of formal initial segments turns out to be the Heyting algebra of open subsets of a compact space Spp($\mathbb{T}$) to be called the spectrum of the theory $\mathbb{T}$. Some results of chapter 2 give rise to a sheaf representation theorem for $\mathbb{T}$-algebras on this spectrum Spp($\mathbb{T}$).

Chapter 6 is devoted to the case of the theory of modules on an arbitrary ring R with a unit. We establish an isomorphism between the frame of formal initial segments of $\underline{\text{Mod}}_R$ and the frame of pure ideals of the ring R. Applying the results of chapter 5, we present R as the ring of global sections of a sheaf of rings on the spectrum of the theory of R-modules; the functorial description of this sheaf is the sheaf of rings of R-linear endomorphisms of the pure ideals of R. An analogous representation theorem holds for any R-module. By interpreting the results of chapter 3, we also obtain the classification of localizations of $\underline{\text{Mod}}_R$ as presented by H. Simmons in [22].

All the material of chapter 6 concerning pure ideals and the sheaf representation theorem has been obtained as a corollary of the general theory developed previous chapters. We have found it interesting to rewrite these results using only standard techniques of ring theory. This is the object of chapter 7 which thus provides, in the very special case of modules, an approach of the representation theorem which becomes independant of the general categorical machinery. We include also some additional results and in particular an alternative representation theorem on the same spectrum of the theory of R-modules.

Finally, in chapter 8, we turn to the case of a Gelfand ring R. We first prove some useful properties of pure ideals in Gelfand rings and also some characterizations of Gelfand rings in terms of pure ideals. This allows us to prove that the sheaf representation of chapters 5 - 6 - 7, in the case of Gelfand rings, is the representation studied by Mulvey and Bkouche in [16] and [3]; in particular it coincides with Pierce's representation in the case of a Von Neumann regular ring. As a consequence, we obtain a functorial description of the classical sheaf representations for Gelfand and Von Neumann regular rings : this is simply the sheaf of R-linear endomorphisms of the pure ideals of the ring R. In an appendix, we show that this description in terms of rings of endomorphisms holds in fact for any ring in the case of Pierce's representation.

We are indebted to Harold Simmons for translating in the non commutative case several of our proofs on commutative Gelfand rings. This work has also been improved by fruitful conversations with M. Carral, C. Mulvey and the participants of the category seminar in Louvain-la-Neuve. This is the opportunity for us to thank all of them.

CONTENTS

CHAPTER 1 : CATEGORIES OF ALGEBRAIC SHEAVES

This chapter does not present any new results, except some technical lemmas which will be useful later. We recall some standard facts on sheaves and algebraic theories and take the opportunity to set out the notations and the terminology.

§ 1. ALGEBRAIC THEORIES

A classical or external finitary algebraic theory $\mathbb{T}$ can be presented as a category with a countable set of distinct objects T^0, T^1, T^2, ..., T^n, ... such that T^n is the n-th power of T^1. A (classical) model of $\mathbb{T}$ is a finite product preserving covariant functor from $\mathbb{T}$ to the category Sets of sets; such a model is also called a (classical) $\mathbb{T}$-algebra. A morphism between two $\mathbb{T}$-algebras is simply a natural transformation. We denote by $\underline{\mathrm{Sets}}^{\mathbb{T}}$ the category of $\mathbb{T}$-algebras and their morphisms. There is a forgetful functor $U : \underline{\mathrm{Sets}}^{\mathbb{T}} \to \underline{\mathrm{Sets}}$ which sends a $\mathbb{T}$-algebra A to the underlying set $A(T^1)$. U has a monomorphism preserving left adjoint $F : \underline{\mathrm{Sets}} \to \underline{\mathrm{Sets}}^{\mathbb{T}}$. F is such that for any finite set n, F(n) is isomorphic to $\mathbb{T}(T^n,-)$; so the set underlying F(n) is the set of n-ary operations.

The category $\underline{\mathrm{Sets}}^{\mathbb{T}}$ is complete and cocomplete. The forgetful functor U preserves and reflects limits and filtered colimits; it is represented by the generator $F(1) \cong \mathbb{T}(T^1,-)$ and thus is faithful. A filtered colimit $L = \varinjlim A_i$ is just the set of all elements in all the A_i divided by the equivalence relation which identifies $x \in A_i$ and $y \in A_j$ if there are morphisms $A_i \to A_k$ and $A_j \to A_k$ which send x and y to the same $z \in A_k$. From this it follows that in $\underline{\mathrm{Sets}}^{\mathbb{T}}$, finite limits commute with filtered colimits. It is also the case that a morphism f in $\underline{\mathrm{Sets}}^{\mathbb{T}}$ is a coequalizer if and only if U(f) is a surjection. Moreover any $\mathbb{T}$-algebra is a quotient of a free $\mathbb{T}$-algebra, i.e. for any $\mathbb{T}$-algebra A there exists a set S and a coequalizer $F(E) \to A$; in fact, E can be chosen to be the underlying set of A.

If $\mathbb{T}$ and $\mathbb{T}'$ are two algebraic theories, a morphism of theories $\mathbb{T} \to \mathbb{T}'$ is a product preserving functor. This induces by composition an algebraic functor $\underline{\mathrm{Sets}}^{\mathbb{T}'} \to \underline{\mathrm{Sets}}^{\mathbb{T}}$; this functor has a left adjoint. It should be noted that a morphism of theories takes any n-ary operation of $\mathbb{T}$ into a n-ary operation of $\mathbb{T}'$.

The results already mentioned can be found in [21], chapter 18. The following facts on commutative theories can be found in [15]. The theory $\mathbb{T}$ is called commutative if for any integers n, m and any operations $\alpha : T^n \to T^1$, $\beta : T^m \to T^1$ the following square commutes :

$$\begin{array}{ccc}
T^{n\times m} \cong (T^n)^m & \xrightarrow{\alpha^m} & T^m \\
\wr\Vert & & \\
(T^m)^n & \hookrightarrow & \\
\beta^n \downarrow & & \downarrow \beta \\
T^n & \xrightarrow[\alpha]{} & T^1
\end{array}$$

when $\mathbb{T}$ is commutative, $\underline{\text{Sets}}^{\mathbb{T}}$ becomes in a natural way a symmetric monoidal closed category.

§ 2. FRAMES

A lattice $\mathbb{H}$ is a partially ordered set in which each pair (u,v) of elements has an infimum $u \wedge v$ and a supremum $u \vee v$. The lattice $\mathbb{H}$ is distributive if for any elements u, v, w of $\mathbb{H}$ the following equalities hold

$$u \wedge (v \vee w) = (u \wedge v) \vee (u \wedge w)$$
$$u \vee (v \wedge w) = (u \vee v) \wedge (u \vee w);$$

in fact each of these equalities implies the other one. The lattice $\mathbb{H}$ is a Heyting algebra if it possesses a smallest element 0, a greatest element 1 and if for any v, w in $\mathbb{H}$ there exists some (necessarily unique) $v \Rightarrow w$ in $\mathbb{H}$ such that for any u in $\mathbb{H}$

$$u \wedge v \leqslant w \quad \text{iff} \quad u \leqslant v \Rightarrow w;$$

a Heyting algebra is automatically a distributive lattice.

A lattice $\mathbb{H}$ is called complete if each subset of $\mathbb{H}$ has a supremum or, equivalently, if each subset of $\mathbb{H}$ has an infimum. A frame is a complete lattice which satisfies the generalized distributive law

$$u \wedge (\bigvee_{i\in I} v_i) = \bigvee_{i\in I} (u \wedge v_i).$$

A frame is necessarily a distributive lattice but the distributive law

$$u \vee (\bigwedge_{i\in I} v_i) = \bigwedge_{i\in I} (u \vee v_i)$$

holds only for finite I. If $\mathbb{H}$ and $\mathbb{H}'$ are two frames, a morphism of frames $f : \mathbb{H} \to \mathbb{H}'$ is a map $f : \mathbb{H} \to \mathbb{H}'$ preserving finite $\wedge$ and arbitrary $\vee$. The notion of frame is equivalent to that of complete Heyting algebra. A morphism of frames does not preserve the "implication" $v \Rightarrow w$.

If X is a topological space, the lattice of open subsets of X is a frame for the usual laws of intersection and union. If $f : X \to Y$ is a continuous mapping between two spaces, f induces by inverse image a morphism of frames $\mathcal{O}(f) : \mathcal{O}(Y) \to \mathcal{O}(X)$ between the corresponding lattices of open subsets. This gives rise to a contravariant functor from the category of topological spaces to the category of frames; this functor has an adjoint which takes a frame into a sober space (i.e. a space such that any closed subset which is not expressible as the union of two proper closed subsets is the closure of exactly one point). All the material we need concerning lattices and frames can be found in [11].

§ 3. SHEAVES ON A FRAME

A frame $\mathbb{H}$ can be seen as a category whose objects are the points of $\mathbb{H}$; there is a (single) morphism from u to v if $u \leqslant v$. A presheaf on $\mathbb{H}$ is a contravariant functor $A : \mathbb{H}^{op} \to \underline{Sets}$; a morphism of presheaves is a natural transformation. If $u \leqslant v$ in $\mathbb{H}$, A is a presheaf on $\mathbb{H}$ and x an element in A(v), we denote by $x|_u$ the image of x in A(u) under the map $A(u \leqslant v)$. The category of presheaves on $\mathbb{H}$ is denoted by $Pr(\mathbb{H})$. A presheaf A is called separated if for any $u = \bigvee_{i\in I} u_i$ in $\mathbb{H}$ and x, y in A(u),

$$x = y \quad \text{iff} \quad \forall\, i \in I \;\; x|_{u_i} = y|_{u_i}.$$

A presheaf A is called a sheaf if for any $u = \bigvee_{i\in I} u_i$ in $\mathbb{H}$ and x_i in $A(u_i)$, the condition

$$\forall\, i, j \in I \qquad x_i|_{u_i \wedge u_j} = x_j|_{u_i \wedge u_j}$$

implies the existence of a unique x in A(u) such that for any i, $x|_{u_i} = x_i$. A sheaf is necessarily separated. The full subcategory of sheaves is denoted by $Sh(\mathbb{H})$; the canonical inclusion $Sh(\mathbb{H}) \hookrightarrow Pr(\mathbb{H})$ has a left adjoint which preserves finite limits : it is called the associated sheaf functor and denoted by $a : Pr(\mathbb{H}) \to Sh(\mathbb{H})$. (Cf. [1]). If A is a separated presheaf and $u \in \mathbb{H}$, aA(u) has an easy description : consider all the families $(x_i \in A(u_i))_{i\in I}$ for all the coverings $u = \bigvee_{i\in I} u_i$ in $\mathbb{H}$, such that

$$\forall\, i, j \in I \qquad x_i|_{u_i \wedge u_j} = x_j|_{u_i \wedge u_j} ;$$

two such families are equivalent if they coincide on all the elements of a common refinement of the coverings; aA(u) is the quotient by the equivalence relation of the set of all such families. In that case the canonical morphism $A \to aA$ is a monomorphism.

§ 4. ALGEBRAIC SHEAVES (EXTERNAL VERSION)

If $\mathbb{T}$ is a (classical) finitary algebraic theory and $\mathbb{H}$ is a frame, a presheaf of $\mathbb{T}$-algebras is a contravariant functor $A : \mathbb{H}^{op} \to \underline{\text{Sets}}^{\mathbb{T}}$; a morphism of presheaves of $\mathbb{T}$-algebras is a natural transformation. The corresponding category is denoted by $\text{Pr}(\mathbb{H}, \mathbb{T})$. There is a forgetful functor $U : \text{Pr}(\mathbb{H}, \mathbb{T}) \to \text{Pr}(\mathbb{H})$ obtained by composition with the forgetful functor $U : \underline{\text{Sets}}^{\mathbb{T}} \to \underline{\text{Sets}}$. U has a left adjoint F preserving monomorphisms and such that for any presheaf $A : \mathbb{H} \to \underline{\text{Sets}}$, $FA(u)$ is the free $\mathbb{T}$-algebra on $A(u)$. A sheaf of $\mathbb{T}$-algebras is a presheaf of $\mathbb{T}$-algebras whose underlying presheaf is a sheaf. The corresponding category of sheaves of $\mathbb{T}$-algebras is denoted by $\text{Sh}(\mathbb{H}, \mathbb{T})$. The canonical full inclusion $\text{Sh}(\mathbb{H}, \mathbb{T}) \hookrightarrow \text{Pr}(\mathbb{H}, \mathbb{T})$ has a left adjoint which preserves finite limits; the reflection of a presheaf of $\mathbb{T}$-algebras is the sheaf universally associated to the underlying presheaf. As a consequence there is a forgetful functor $U : \text{Sh}(\mathbb{H}, \mathbb{T}) \to \text{Sh}(\mathbb{H})$ which has a monomorphism preserving left adjoint sending a sheaf A to $aF(A)$. All these results on sheaves can be found in [1].

§ 5. ALGEBRAIC SHEAVES (INTERNAL VERSION)

If $\mathbb{H}$ is a frame, $\text{Sh}(\mathbb{H})$ is a topos satisfying the axiom of infinity and it makes sense to speak of a finitary algebraic theory $\mathbb{T}$ internally defined with respect to $\text{Sh}(\mathbb{H})$. This is exactly a sheaf on $\mathbb{H}$ with values in the category of algebraic theories and their morphisms. In other words, $\mathbb{T}$ is a contravariant functor from $\mathbb{H}$ to the category of algebraic theories and their morphisms, such that that for any integer n, the functor $\mathbb{H}^{op} \to \underline{\text{Sets}}$ which sends $u \in \mathbb{H}$ to the set $\mathcal{O}_n(u)$ of n-ary operations of the theory $\mathbb{T}(u)$ is a sheaf in the usual sense.

A $\mathbb{T}$-algebra in $\text{Sh}(\mathbb{H})$ is a sheaf $A : \mathbb{H}^{op} \to \underline{\text{Sets}}$ equipped, for any $u \in \mathbb{H}$, with the structure of a $\mathbb{T}(u)$ algebra on $A(u)$ in such a way that for $u \leqslant v$ in $\mathbb{H}$ and $\alpha \in \mathcal{O}_n(v)$ the following diagram commutes

$$\begin{array}{ccc} A^n(v) & \xrightarrow{\quad\alpha\quad} & A(v) \\ {\scriptstyle A^n(u \leqslant v)}\Big\downarrow & \hookrightarrow & \Big\downarrow{\scriptstyle A(u \leqslant v)} \\ A^n(u) & \xrightarrow[\quad\alpha|_u\quad]{} & A(u). \end{array}$$

A morphism $f : A \to B$ of $\mathbb{T}$-algebras in Sh($\mathbb{H}$) is a natural transformation such that for any $u \in \mathbb{H}$, f_u is a morphism of $\mathbb{T}(u)$-algebras. The category of $\mathbb{T}$-algebras in Sh($\mathbb{H}$) is denoted by Sh($\mathbb{H}$, $\mathbb{T}$). An analogous definition holds for presheaves and we get a category Pr($\mathbb{H}$, $\mathbb{T}$).

Sh($\mathbb{H}$, $\mathbb{T}$) is a full subcategory of Pr($\mathbb{H}$, $\mathbb{T}$) and the canonical inclusion has a left adjoint a which preserves finite limits and is the associated sheaf functor. Moreover the obvious forgetful functor U : Pr($\mathbb{H}$, $\mathbb{T}$) $\to$ Pr($\mathbb{H}$) has a monomorphism preserving left adjoint F : Pr($\mathbb{H}$) $\to$ Pr($\mathbb{H}$, $\mathbb{T}$) such that, for any presheaf A and any element $u \in \mathbb{H}$, $FA(u)$ is the free $\mathbb{T}(u)$-algebra on $A(u)$. This implies that the forgetful functor U : Sh($\mathbb{H}$, $\mathbb{T}$) $\to$ Sh($\mathbb{H}$) has a monomorphism preserving left adjoint which sends a sheaf A to $aF(A)$. These results on internal algebraic theories can be found in classical texts on topos theory, like [12].

We used the same notation Sh($\mathbb{H}$, $\mathbb{T}$) in both cases of a classical theory $\mathbb{T}$ and a theory internally defined with respect to Sh($\mathbb{H}$). In fact no real confusion arises because the former situation is a special case of the latter as can be seen from the following argument : a classical finitary algebraic theory $\mathbb{T}$ may be identified with a constant presheaf $\Delta\mathbb{T}$ of algebraic theories on $\mathbb{H}$; the corresponding associated sheaf $a\Delta\mathbb{T}$ is a theory internally defined with respect to Sh($\mathbb{H}$) and the categories Sh($\mathbb{H}$, $\mathbb{T}$) and Sh($\mathbb{H}$, $a\Delta\mathbb{T}$) coincide. For this reason we shall work in the more general context of a theory $\mathbb{T}$ internally defined with respect to Sh($\mathbb{H}$).

From now on and through this chapter $\mathbb{H}$ is a frame and $\mathbb{T}$ is a finitary algebraic theory internally defined with respect to Sh($\mathbb{H}$). We recall and establish some basic facts about Sh($\mathbb{H}$, $\mathbb{T}$).

§ 6. LIMITS AND COLIMITS

Proposition 1.

The categories Pr($\mathbb{H}$, $\mathbb{T}$) *and* Sh($\mathbb{H}$, $\mathbb{T}$) *are complete, cocomplete and regular.*

Any algebraic category is complete, cocomplete and regular. Now in Pr($\mathbb{H}$, $\mathbb{T}$) limits, colimits and images are computed pointwise : this implies that Pr($\mathbb{H}$, $\mathbb{T}$) is complete, cocomplete and regular. Sh($\mathbb{H}$, $\mathbb{T}$) is complete and cocomplete as a full reflective subcategory of Pr($\mathbb{H}$, $\mathbb{T}$); it is regular because the reflection is exact. (Cfr. [2]). ■

Proposition 2.

The forgetful functors $U : Pr(H, T) \to Pr(H)$ *and* $U : Sh(H, T) \to Sh(H)$ *preserve and reflect filtered colimits.*

In any algebraic category the filtered colimits are computed as in the category of sets. In $Pr(H, T)$ and $Pr(H)$ all colimits are computed pointwise. Therefore the filtered colimits in $Pr(H, T)$ are computed as in $Pr(H)$.

To compute an arbitrary colimit in $Sh(H, T)$ or in $Pr(H, T)$, we need to compute it in $Pr(H, T)$ or $Pr(H)$ and apply the associated sheaf functor. But filtered colimits are computed in the same way in $Pr(H, T)$ and $Pr(H)$ and the associated sheaf functor preserves them. So the result holds in the case of sheaves. ■

Proposition 3.

In $Pr(H, T)$ *and* $Sh(H, T)$, *finite limits commute with filtered colimits.*

This is true in any algebraic category and hence it is in $Pr(H, T)$ where limits and colimits are computed pointwise. In $Sh(H, T)$, a limit or a colimit is the reflection of the corresponding limit or colimit in $Pr(H, T)$; as the reflection preserves colimits and finite limits, the commutation property transfers to $Sh(H, T)$. ■

If u is some element in H, we denote by $h_u : H^{op} \to \underline{Sets}$ the presheaf represented by u; the continuity of a representable functor implies immediately that h_u is in fact a sheaf.

§ 7. ALGEBRAIC YONEDA LEMMAS

Proposition 4.

Consider $u \in H$ *and* $A \in Pr(H, T)$. *The following natural isomorphism holds*

$$A(u) \cong (F\, h_u, A).$$

$$\begin{aligned} A(u) &\cong UA(u) && \\ &\cong (h_u, UA) && \text{Yoneda lemma} \\ &\cong (F\, h_u, A) && \text{adjunction } F \dashv U. \end{aligned}$$ ■

Proposition 5.

Consider $u \in \mathbf{H}$ *and* $A \in Sh(\mathbf{H}, \mathbb{T})$. *The following natural isomorphism holds*

$$A(u) \cong (a\,F\,h_u, A).$$

$$\begin{aligned} A(u) &\cong (F\,h_u, A) && \text{proposition 9} \\ &\cong (a\,F\,h_u, A) && \text{adjunction.} \end{aligned}$$ ■

§ 8. GENERATORS

Proposition 6.

The set of presheaves $F\,h_u$ *with* $u \in \mathbf{H}$ *is a set of finitely presentable regular generators in* $Pr(\mathbf{H}, \mathbb{T})$. *The set of sheaves* $a\,F\,h_u$ *with* $u \in \mathbf{H}$ *is a set of regular generators in* $Sh(\mathbf{H}, \mathbb{T})$.

Let us first consider the case of presheaves. If $f \neq g : A \rightrightarrows B$ are two different arrows in $Pr(\mathbf{H}, \mathbb{T})$, there is some element $u \in \mathbf{H}$ such that $f_u \neq g_u : A(u) \rightrightarrows B(u)$. Hence, there is some x in $A(u)$ such that $f_u(x) \neq g_u(x)$. By proposition 4 this produces a morphism $x' : F\,h_u \to A$ in $Pr(\mathbf{H}, \mathbb{T})$ such that $f \circ x' \neq g \circ x'$. This shows that the $F\,h_u$ are a family of generators.

In fact the $F\,h_u$ are a proper set of generators. Indeed, consider a morphism $f : A \to B$ in $Pr(\mathbf{H}, \mathbb{T})$ such that for any $u \in \mathbf{H}$ the map

$$(1, f) : (F\,h_u, A) \to (F\,h_u, B)$$

is an isomorphism; we have to show that f is an isomorphism (cfr. [7]). By adjunction the map

$$(1, Uf) : (h_u, UA) \to (h_u, UB)$$

is an isomorphism and by the Yoneda lemma the map

$$(Uf)_a : UA(a) \to UB(a)$$

is an isomorphism. Thus f is pointwise a bijection and thus pointwise an isomorphism; therefore f is an isomorphism.

By [7] - 1 - 7 ((ii) ⇒ (iii)) any epimorphism in $Pr(\mathbf{H}, \mathbb{T})$ is regular. By [7] - 1 - 10, for any object $A \in Pr(\mathbf{H}, \mathbb{T})$ there is a proper, and therefore regular, epimorphism of the form

$$p : \coprod_i F\,h_{u_i} \twoheadrightarrow A$$

where I is some indexing set. Now consider the canonical epimorphism

$$q : \coprod_{(u,g)} F\,h_u \to A \qquad (g : F\,h_u \to A),$$

p factors through q into a map r $(q \circ r = p)$ defined by $r \circ s_{u_i} = s_{(u_i,\, p \circ s_{u_i})}$.

Hence, q is a regular epimorphism and $\{F\,h_u \mid u \in \mathbb{H}\}$ is a regular set of generators in $Pr(\mathbb{H}, \mathbb{T})$.

Now if $A = \varinjlim A_i$ is a filtered colimit in $Pr(\mathbb{H}, \mathbb{T})$, the following isomorphisms hold

$$
\begin{array}{lll}
(F\,h_u, A) & & \\
\quad \cong (F\,h_u, \varinjlim A_i) & & \\
\quad \cong (h_u, U \varinjlim A_i) & & \text{by adjunction} \\
\quad \cong (h_u, \varinjlim UA_i) & & U \text{ preserves filtered colimits} \\
\quad \cong (\varinjlim UA_i)(u) & & \text{Yoneda lemma} \\
\quad \cong \varinjlim (UA_i)(u) & & \text{colimits are pointwise} \\
\quad \cong \varinjlim (h_u, UA_i) & & \text{Yoneda lemma} \\
\quad \cong \varinjlim (F\,h_u, A_i) & & \text{by adjunction.}
\end{array}
$$

This shows that $F\,h_u$ is finitely presentable (cfr. [7]).

Now if A is some object in $Sh(\mathbb{H}, \mathbb{T})$, consider the canonical morphism

$$p : \coprod_{u,f} a\,F\,h_u \to A \qquad (f : a\,F\,h_u \to A)$$

in $Sh(\mathbb{H}, \mathbb{T})$ and

$$q : \coprod_{u,g} F\,h_u \to A \qquad (g : F\,h_u \to A)$$

in $Pr(\mathbb{H}, \mathbb{T})$. By adjunction there is an isomorphism

$$(a\,F\,h_u, A) \cong (F\,h_u, A)$$

and therefore the following diagram commutes in $Pr(\mathbb{H}, \mathbb{T})$

$$
\begin{array}{ccc}
\coprod_{u,f} F\,h_u & \xrightarrow{\ \alpha\ } & a(\coprod_{u,f} F\,h_u) \\
\wr\Vert & & \wr\Vert \\
\coprod_{u,g} F\,h_u & \hookrightarrow & \coprod_{u,f} a\,F\,h_u \\
 & {\scriptstyle q}\searrow \quad \swarrow{\scriptstyle p} & \\
 & A &
\end{array}
$$

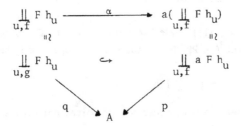

(the coproduct $\coprod_{u,f} a\, F\, h_u$ is computed in Sh(H, T), the other ones are computed in Pr(H, T); α is the canonical morphism arising from the adjunction). Applying the associated sheaf functor to this diagram, it appears that p is isomorphic to a(q); therefore p is a regular epimorphism, which concludes the proof. ∎

Observe that it is not true in general that the objects $a\, F\, h_u$ of Sh(H, T) are finitely presentable. The reason for this is that filtered colimits in Sh(H, T) are not computed pointwise.

Proposition 7.

The categories Pr(H, T) *and* Sh(H, T) *have a dense family of generators* $(G_i)_{i\in I}$ *whose elements are such that the unique morphism* $0 \to G_i$ *is a monomorphism.*

We know by [7] (7 - 5) that the finite sums of the generators $F\, h_u$ in Pr(H, T) are a dense family of generators. But in Pr(H), 0 is the constant functor on the empty set; so 0 is a strict initial object (any morphism with codomain 0 is an isomorphism); therefore any morphism with domain 0 in Pr(H) is a monomorphism. Now consider $\coprod_{i\in I} F\, h_{u_i}$ in Pr(H, T), for any indexing set I. The morphism

$$F(0) \cong 0 \to \coprod_{i\in I} F\, h_{u_i} \cong F(\coprod_{i\in I} h_{u_i})$$

is the image of the monomorphism

$$0 \to \coprod_{i\in I} h_{u_i}$$

in Pr(H); as F preserves monomorphisms, this is a monomorphism. This proves the result for Pr(H, T).

Now let A be some object in Sh(H, T). Looking at A as an object in Pr(H, T) we can write

$$A = \varinjlim_{g} \coprod_{i\in I} F\, h_{u_i}$$

where I runs through the finite sets and g is a morphism $g : \coprod_{i\in I} F\, h_{u_i} \to A$.

(Cfr. [7], 3 - 1). Applying the associated sheaf functor, we get

$$A = \varinjlim_{g} \coprod_{i\in I} a\, F\, h_{u_i}.$$

In fact we need to prove that

$$A = \varinjlim_{f} \coprod_{i \in I} a\,F\,h_{u_i}$$

where I runs through the finite sets and f is a morphism $f : \coprod_{i \in I} a\,F\,h_{u_i} \to A$.

In fact, by adjunction, the morphisms f and g are in a one-to-one correspondance. So the difference between both colimits is that in the first case, the diagram defining the colimit contains only morphisms of the form $a(k) : (\coprod_{i \in I} a\,F\,h_{u_i}, g_1) \to (\coprod_{j \in J} a\,F\,h_{u_j}, g_2)$ where k makes the following diagram commute

$$\begin{array}{ccc} \coprod_{i \in I} F\,h_{u_i} & \xrightarrow{\;g_1\;} & \\ {\scriptstyle k}\downarrow & \circlearrowright & A \\ \coprod_{j} F\,h_{u_j} & \xrightarrow{\;g_2\;} & \end{array} \quad ;$$

however in the second case the diagram defining the colimit contains all the morphisms $\ell : (\coprod_{i \in I} a\,F\,h_{u_i}, f_1) \to (\coprod_{j \in J} a\,F\,h_{u_j}, f_2)$ where ℓ makes the following diagram commute

$$\begin{array}{ccc} \coprod_{i \in I} a\,F\,h_{u_i} & \xrightarrow{\;f_1\;} & \\ {\scriptstyle \ell}\downarrow & \circlearrowright & A \\ \coprod_{j \in J} a\,F\,h_{u_j} & \xrightarrow{\;f_2\;} & \end{array} \quad .$$

In fact this does not change anything in the computation of the colimit. Indeed, both diagrams have the same objects and the colimit cone on the small diagram is, by definition of the big diagram, a cone on this big diagram. Now any cone on the big diagram induces a cone on the small diagram and thus a unique factorization. So both colimits coincide. This proves that the $a\,F\,h_u$ form a dense family of generators in $Sh(\mathbb{H}, \mathbb{T})$.

Now $0 \to \coprod_{i \in I} F\,h_{u_i}$ is a monomorphism. Applying the associated sheaf functor, which is exact, we get a monomorphism $0 \to \coprod_{i \in I} a\,F\,h_{u_i}$ as required. ∎

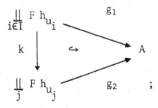

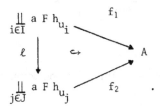

§ 9. FILTERED UNIONS

Proposition 8.

In $\mathrm{Pr}(\mathbb{H}, \mathbb{T})$ *and* $\mathrm{Sh}(\mathbb{H}, \mathbb{T})$, *intersection with a fixed subobject commutes with filtered unions of subobjects.*

This fact is true in any algebraic category (cfr. [21] - 18 - 3 - 7) and thus holds in $\mathrm{Pr}(\mathbb{H}, \mathbb{T})$ where unions and intersections are computed pointwise.

The associated sheaf functor beeing exact commutes with intersections and unions. Therefore the theorem holds in $\mathrm{Sh}(\mathbb{H}, \mathbb{T})$. ■

Proposition 9.

In $\mathrm{Pr}(\mathbb{H}, \mathbb{T})$ *and* $\mathrm{Sh}(\mathbb{H}, \mathbb{T})$ *the following facts hold. Let* $(A_i \rightarrowtail A)_{i\in I}$ *be a family of subobjects and* $(f_i : A_i \to B)_{i\in I}$ *a family of morphisms such that for any indexes* i, j *in* I, *the following diagram commutes :*

$$\begin{array}{ccc} A_i \cap A_j & \hookrightarrow & A_j \\ \downarrow & \hookrightarrow & \downarrow f_j \\ A_i & \xrightarrow[f_i]{} & B. \end{array}$$

If the family $(A_i)_{i\in I}$ *of subobjects is filtered, there exists a unique morphism* $f : \bigcup_{i\in I} A_i \to B$ *extending all the* f_i*'s; moreover if each* f_i *is a monomorphism, the same holds for* f.

By exactness of the associated sheaf functor, it is again sufficient to check the result in $\mathrm{Pr}(\mathbb{H}, \mathbb{T})$. But in $\mathrm{Pr}(\mathbb{H}, \mathbb{T})$ intersections, unions and monomorphisms are described pointwise. So it is sufficient to prove the result for an algebraic category. But in an algebraic category intersections and filtered unions are computed as in the category of sets (cfr. [21] - 18 - 3). Therefore it is sufficient to check the result in the category of sets where it holds. ■

Proposition 10.

In $\mathrm{Pr}(\mathbb{H}, \mathbb{T})$ *and* $\mathrm{Sh}(\mathbb{H}, \mathbb{T})$, *a filtered colimit of monomorphisms is a monomorphism.*

Again this is true in every algebraic category (cfr. [21] - 18 - 3 - 7) and thus in Pr(H, T) by a pointwise argument and in Sh(H, T) by the exactness of the associated sheaf functor. ■

CHAPTER 2 : THE FORMAL INITIAL SEGMENTS

Throughout this chapter, $\mathbb{H}$ is a frame and $\mathbb{T}$ is an algebraic theory internally defined with respect to the topos $Sh(\mathbb{H})$ of sheaves on $\mathbb{H}$.

If u is some element in $\mathbb{H}$, we consider the initial segment induced by u in $\mathbb{H}$

$$u{\downarrow} = \{v \in \mathbb{H} \mid v \leqslant u\};$$

$u{\downarrow}$ is also a frame. $\mathbb{T}$ is a sheaf of algebraic theories on $\mathbb{H}$ and, by restricting this sheaf to $u{\downarrow}$, we obtain an algebraic theory $\mathbb{T}_{u\downarrow}$ internally defined with respect to the topos $Sh(u{\downarrow})$. In this chapter we compare the categories $Sh(\mathbb{H},\mathbb{T})$ and $Sh(u{\downarrow}, \mathbb{T}_{u\downarrow})$ of algebraic sheaves.

$Sh(u{\downarrow}, \mathbb{T}_{u\downarrow})$ is always a localization of $Sh(\mathbb{H},\mathbb{T})$, but this localization has a lot of additional properties. One of these properties is expressed in terms of what we call "Heyting subobjects". In a topos, it is well known that the lattice of subobjects of a fixed object is in fact a Heyting algebra. In $Sh(\mathbb{H},\mathbb{T})$, this is no longer true : the distributivity conditions are lacking as well as some properties of the union of subobjects. However, for some convenient subobjects, these properties still hold in $Sh(\mathbb{H},\mathbb{T})$: these subobjects are called Heyting subobjects; they will be important in our study. The set of Heyting subobjects of some fixed object in $Sh(\mathbb{H},\mathbb{T})$ is a frame.

We list some of the basic properties of $Sh(u{\downarrow}, \mathbb{T}_{u\downarrow})$ with respect to $Sh(\mathbb{H},\mathbb{T})$. A localization of $Sh(\mathbb{H},\mathbb{T})$ which satisfies these properties is called a formal initial segment of the category $Sh(\mathbb{H},\mathbb{T})$. In chapter 4, we shall establish, for a convenient theory $\mathbb{T}$, a one-to-one correspondance between the initial segments of $\mathbb{H}$ and the formal initial segments of $Sh(\mathbb{H},\mathbb{T})$. But this bijection does not hold in general.

The main result of this chapter shows that the formal initial segments of $Sh(\mathbb{H},\mathbb{T})$ constitute a frame which contains $\mathbb{H}$ as a subframe and is itself contained as a subframe in the frame of Heyting subobjects of the free algebra $F\,h_1$ in $Sh(\mathbb{H},\mathbb{T})$.

Finally, when the theory $\mathbb{T}$ is defined externally (i.e. when $\mathbb{T}$ is the sheaf associated to a constant presheaf), we compare $Sh(\mathbb{H},\mathbb{T})$ with the category of sheaves of $\mathbb{T}$-algebras on the frame of formal initial segments of $Sh(\mathbb{H},\mathbb{T})$: the restriction functor has a left exact left adjoint.

§ 1. HEYTING SUBOBJECTS

We first define the notion of a Heyting subobject.

Definition 1.

Let C *be a category equivalent to some category* $\mathrm{Sh}(\mathbb{H}, \mathbb{T})$. *A subobject* $R \rightarrowtail A$ *in* C *is called a Heyting subobject of* A *if for any subobjects* S, T *of* A *the following conditions hold :*

(H 1) $\qquad R \cap (S \cup T) = (R \cap S) \cup (R \cap T)$

(H 2) $\qquad S \cap (R \cup T) = (S \cap R) \cup (S \cap T)$

(H 3) *the square*

$$\begin{array}{ccc} R \cap S \cap T & \longrightarrow & S \cap T \\ \downarrow & & \downarrow \\ R \cap T & \longrightarrow & (R \cap T) \cup (S \cap T) \end{array}$$

is cocartesian.

This notion is stable by restriction.

Proposition 1.

Let C *be a category equivalent to some category* $\mathrm{Sh}(\mathbb{H}, \mathbb{T})$. *Let* $R \rightarrowtail A$ *be a Heyting subobject in* C *and* $B \rightarrowtail A$ *any subobject. Then* $R \cap B$ *is a Heyting subobject of* B.

Let S, T be any subobjects of B.

$$\begin{array}{rlr} (R \cap B) \cap (S \cup T) & = R \cap (S \cup T) & S \cup T \subseteq B \\ & = (R \cap S) \cup (R \cap T) & H\ 1 \\ & = ((R \cap B) \cap S) \cup ((R \cap B) \cap T) & S \subseteq B;\ T \subseteq B \end{array}$$

$$\begin{array}{rlr} S \cap ((R \cap B) \cup T) & = S \cap ((R \cup T) \cap (B \cup T)) & H\ 2 \\ & = S \cap (R \cup T) \cap B & T \subseteq B \\ & = S \cap (R \cup T) & S \subseteq B \\ & = (S \cap R) \cup (S \cap T) & H\ 2 \\ & = (S \cap (R \cap B)) \cup (S \cap T) & S \subseteq B. \end{array}$$

Finally the square

$$\begin{array}{ccc} (R \cap B) \cap S \cap T & \longrightarrow & S \cap T \\ \downarrow & & \downarrow \\ (R \cap B) \cap T & \longrightarrow & ((R \cap B) \cap T) \cup (S \cap T) \end{array}$$

is simply the square

$$\begin{array}{ccc} R \cap S \cap T & \longrightarrow & S \cap T \\ \downarrow & & \downarrow \\ R \cap T & \longrightarrow & (R \cap T) \cup (S \cap T) \end{array}$$

because $S \subseteq B$, $T \subseteq B$; it is thus cocartesian. ■

Conditions H 1 and H 2 are finite; but the properties of filtered unions in Sh($\mathbb{H}, \mathbb{T}$) allow us to deduce from them the infinite version of H 1 and H 2.

Proposition 2.

Let C *be a category equivalent to some category* Sh($\mathbb{H}, \mathbb{T}$). *Let* R *and* $(R_i)_{i \in I}$ *be Heyting subobjects of* A *in* C. *Let* S, T *and* $(T_i)_{i \in I}$ *be any subobjects of* A. *The following conditions hold :*

(I H 1) $\quad R \cap (\bigcup_{i \in I} T_i) = \bigcup_{i \in I} (R \cap T_i)$

(I H 2) $\quad S \cap (T \cup (\bigcup_{i \in I} R_i)) = (S \cap T) \cup (\bigcup_{i \in I} (S \cap R_i))$.

An iterated application of H 1 and H 2 shows that proposition 2 holds for a finite indexing set I. For an arbitrary I, let us denote by $F(I)$ the set of finite subsets of I.

$$\begin{aligned} R \cap (\bigcup_{i \in I} T_i) &= R \cap (\bigcup_{J \in F(I)} (\bigcup_{i \in J} T_i)) \\ &= \bigcup_{J \in F(I)} (R \cap (\bigcup_{i \in J} T_i)) && \text{prop. I - 8} \\ &= \bigcup_{J \in F(I)} \bigcup_{i \in J} (R \cap T_i) && \text{J finite} \\ &= \bigcup_{i \in I} (R \cap T_i). \end{aligned}$$

$$S \cap (T \cup (\bigcup_{i\in I} R_i)) = S \cap (T \cup (\bigcup_{J\in F(I)} (\bigcup_{i\in J} R_i)))$$

$$= S \cap (\bigcup_{J\in F(I)} (T \cup (\bigcup_{i\in J} R_i)))$$

$$= \bigcup_{J\in F(I)} (S \cap (T \cup (\bigcup_{i\in J} R_i))) \qquad \text{prop. I - 8}$$

$$= \bigcup_{J\in F(I)} ((S \cap T) \cup (\bigcup_{i\in J} (S \cap R_i))) \qquad \text{J finite}$$

$$= (S \cap T) \cup (\bigcup_{i\in I} (S \cap R_i)).$$

■

The following finite conditions are also valid for Heyting subobjects, which completes the analogy with Heyting algebras and subobjects in a topos.

Proposition 3.

Let C *be a category equivalent to some category* Sh(H, T). *Let* $R \rightarrowtail A$ *be a Heyting subobject in* C. *For any subobjects* S, T *of* A *the following conditions hold* :

(H 4) $R \cup (S \cap T) = (R \cup S) \cap (R \cup T)$

(H 5) $S \cup (R \cap T) = (S \cup R) \cap (S \cup T)$

(H 6) *there exists a subobject* $R \Rightarrow S$ *of* A *such that, for any subobject* Q *of* A

$$Q \leqslant (R \Rightarrow S) \quad \text{iff} \quad Q \cap R \leqslant S.$$

$$(R \cup S) \cap (R \cup T) = ((R \cup S) \cap R) \cup ((R \cup S) \cap T) \qquad \text{H 2}$$
$$= R \cup ((R \cap T) \cup (S \cap T)) \qquad \text{H 2}$$
$$= R \cup (S \cap T).$$

$$(S \cup R) \cap (S \cup T) = (S \cap (S \cup T)) \cup (R \cap (S \cup T)) \qquad \text{H 2}$$
$$= S \cup ((R \cap S) \cup (R \cap T)) \qquad \text{H 1}$$
$$= S \cup (R \cap T).$$

Finally, define

$$(R \Rightarrow S) = \bigcup \{P \subseteq A \mid P \cap R \leqslant S\}.$$

Clearly, if $Q \cap R \leqslant S$, then $Q \leqslant (R \Rightarrow S)$. Conversely, if $Q \leqslant (R \Rightarrow S)$, then

$$Q \cap R \leqslant (R \Rightarrow S) \cap R$$
$$= (\bigcup \{P \subseteq A \mid P \cap R \leqslant S\}) \cap R$$
$$= \bigcup \{P \cap R \mid P \cap R \leqslant S\} \qquad \text{I H 1}$$
$$\leqslant S.$$

■

We now give a proof of the main result of this paragraph, namely that the Heyting subobjects of a fixed object of Sh(H, T) form a frame. In fact, this result holds for more general categories than Sh(H, T), but we do not go into this generalization. We simply state :

Theorem 4.

Let C a category equivalent to some category Sh(H, T). *In C, the Heyting subobjects of a fixed object* A *form a frame for the usual operations* ∩ *and* ∪ *on subobjects of* A.

The conditions H 1 - H 2 - H 3 are obviously satisfied when R = A (notations of definition 1); so the greatest subobject is a Heyting subobject. On the other hand, A has a smallest subobject = the subobject which consists exactly of all constants of A (at each level); let us denote this object by 0_A. For any subobject S of A, $0_A \cap S = 0_A$ and therefore the smallest subobject 0_A is also a Heyting subobject.

In order to get a frame, it is sufficient to show that a finite intersection and an arbitrary union of Heyting subobjects is again a Heyting subobject. Indeed, the infinite distributivity law is already asserted by proposition 2. We prove now these two facts.

Let Q and R be two Heyting subobjects of A in Sh(H, T) and S, T two arbitrary subobjects of A.

$$
\begin{aligned}
(Q \cap R) \cap (S \cup T) &= Q \cap ((R \cap S) \cup (R \cap T)) && \text{H 1}\\
&= (Q \cap R \cap S) \cup (Q \cap R \cap T) && \text{H 1}
\end{aligned}
$$

$$
\begin{aligned}
S \cap ((Q \cap R) \cup T) &= S \cap ((Q \cup T) \cap (R \cup T)) && \text{H 2}\\
&= (S \cap (Q \cup T)) \cap (R \cup T) &&\\
&= ((S \cap Q) \cup (S \cap T)) \cap (R \cup T) && \text{H 2}\\
&= (((S \cap Q) \cup (S \cap T)) \cap R) \cup (((S \cap Q) \cup (S \cap T)) \cap T) && \text{H 2}\\
&= (S \cap Q \cap R) \cup (S \cap T \cap R) \cup (S \cap (Q \cup T) \cap T) && \text{H 1}\\
&= (S \cap Q \cap R) \cup (S \cap T \cap R) \cup (S \cap T) &&\\
&= (S \cap Q \cap R) \cup (S \cap T).
\end{aligned}
$$

Finally consider the following commutative diagram

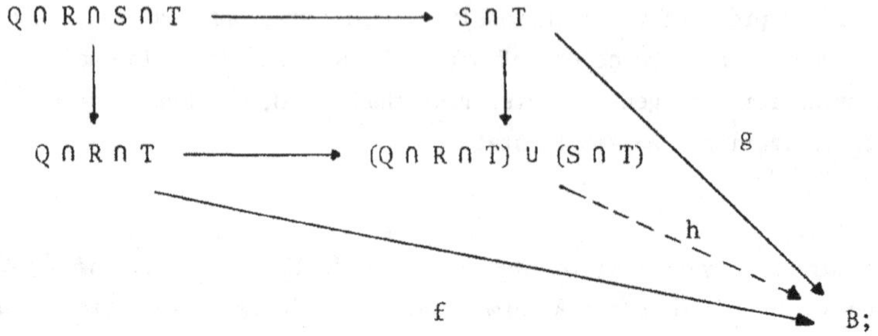

we have to show there exists a unique h making the whole diagram commute. Consider the following diagram where the square is cocartesian and k is the unique extension of f and g.

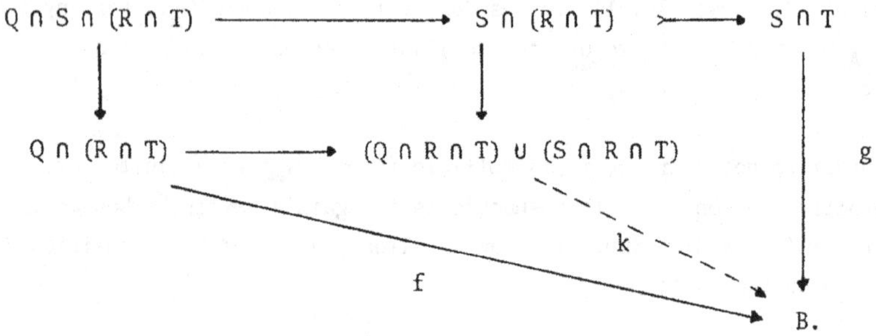

Now by proposition 1 consider $R \cap ((Q \cap T) \cup (S \cap T)) = (Q \cap R \cap T) \cup (R \cap S \cap T)$ (by H 1) as a Heyting subobject of $(Q \cup T) \cap (S \cup T)$. We get a commutative diagram with a cocartesian square and a unique factorization h :

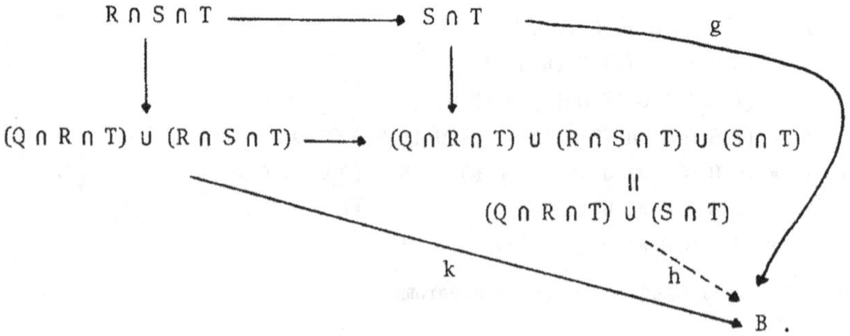

h is clearly the unique morphism extending f and g and this concludes the proof that $Q \cap R$ is a Heyting subobject of A.

Consider now a family $(R_i)_{i \in I}$ of Heyting subobjects of A and two arbitrary subobjects S, T of A. We have to prove that $\bigcup_{i \in I} R_i$ is a Heyting subobject of A.

$$\begin{aligned}
(\bigcup_{i\in I} R_i) \cap (S \cup T) &= \bigcup_{i\in I} (R_i \cap (S \cup T)) && \text{I H 2}\\
&= \bigcup_{i\in I} ((R_i \cap S) \cup (R_i \cap T)) && \text{H 1}\\
&= (\bigcup_{i\in I} (R_i \cap S)) \cup (\bigcup_{i\in I} (R_i \cap T)) && \\
&= ((\bigcup_{i\in I} R_i) \cap S) \cup ((\bigcup_{i\in I} R_i) \cap T) && \text{I H 2}
\end{aligned}$$

$$\begin{aligned}
S \cap ((\bigcup_{i\in I} R_i) \cup T) &= (\bigcup_{i\in I} (S \cap R_i)) \cup (S \cap T) && \text{I H 2}\\
&= (S \cap (\bigcup_{i\in I} R_i)) \cup (S \cap T) && \text{I H 2.}
\end{aligned}$$

In order to prove H 3, consider the following commutative diagram

$$\begin{array}{ccccc}
(\bigcup_{i\in I} R_i) \cap S \cap T & \longrightarrow & S \cap T & & \\
\downarrow & & \downarrow & \searrow^{g} & \\
(\bigcup_{i\in I} R_i) \cap T & \longrightarrow & ((\bigcup_{i\in I} R_i) \cap T) \cup (S \cap T) & & \\
& \searrow_{f} & & & B.
\end{array}$$

By I H 2, we have to show there exists a unique

$$h : ((\bigcup_{i\in I} R_i) \cup S) \cap T \to B$$

extending f and g.

We first produce an extension

$$h_J : ((\bigcup_{i\in J} R_i) \cup S) \cap T \to B$$

for any finite subset J of I. We choose

$$h_\phi : S \cap T \to B$$

to be g. Now if h_J is defined and $j \in I$ we obtain $h_{J\cup\{j\}}$ in the following commutative diagram where the square is cocartesian :

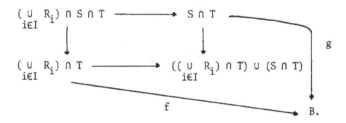

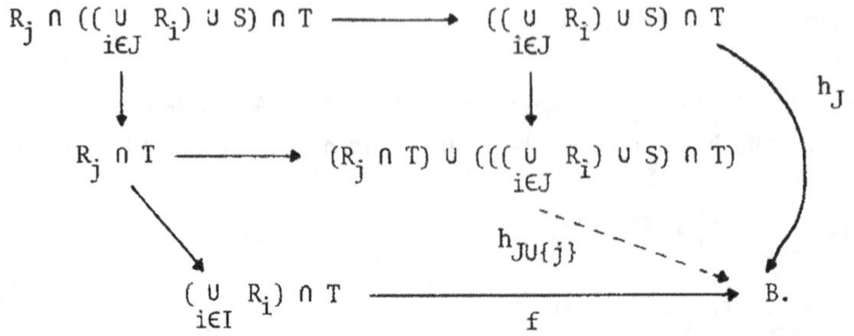

The family $(h_J)_J$ is defined on a filtered family of subobjects and satisfies the conditions of proposition I - 8; therefore we get a unique h extending f and g. ■

§ 2. ALGEBRAIC SHEAVES ON AN INITIAL SEGMENT

Let u be some element in the frame $\mathbb{H}$. We now investigate the relationship between the categories $Sh(\mathbb{H}, \mathbb{T})$ and $Sh(u\downarrow, \mathbb{T}_{u\downarrow})$, where $u\downarrow$ is the initial segment determined by u. As $u\downarrow$ is a subcategory of $\mathbb{H}$ we have a restriction functor

$$u^* : Sh(\mathbb{H}, \mathbb{T}) \to Sh(u\downarrow, \mathbb{T}_{u\downarrow})$$

which acts by composition with the canonical inclusion $u\downarrow \hookrightarrow \mathbb{H}$. In this paragraph, we exhibit several properties of u^*.

We denote by 0 the initial object of $Sh(\mathbb{H}, \mathbb{T})$; it consists at each level of all the constants of the theory $\mathbb{T}$. When each theory $\mathbb{T}(u)$ has no or a single constant (or when $\mathbb{T}$ is the sheaf associated to a presheaf with this property), any morphism $0 \to A$ is necessarily a monomorphism, for in Sets any mapping from the empty set or the singleton is necessarily injective. But in general, when $\mathbb{T}$ has several constants, some of them could be identified by a morphism $0 \to A$ in $Sh(\mathbb{H}, \mathbb{T})$; so this morphism is no longer a monomorphism. This remark takes its meaning in the following theorem.

Theorem 5.

Let u be some element in the frame $\mathbb{H}$ *and* $u^* : Sh(\mathbb{H}, \mathbb{T}) \to Sh(u\downarrow, \mathbb{T}_{u\downarrow})$ *the restriction functor. The following results hold :*

(1) u^* *has a left adjoint* $u_!$ *and a right adjoint* u_*

(2) $u_!$ *and* u_* *are full and faithful.*

(3) $u_!$ *preserves and creates monomorphisms.*

(4) *for any* A *in* $Sh(\mathbb{H}, \mathbb{T})$, *the image of the canonical morphism* $u_! \, u^* A \to A$

arising from the adjunction $u_! \dashv u^*$ *is a Heyting subobject of* A.

(5) *for any monomorphism* $0 \to A$ *in* Sh(H, T), *the canonical morphism* $u_!\, u^*\, A \to A$ *arising from the adjunction* $u_! \dashv u^*$ *is a monomorphism.*

(6) $u_!\, u^*\, (a\, F\, h_1) \cong a\, F\, h_u$.

We start with the definition of u_*. If M is some object in $\mathrm{Sh}(u{\downarrow}, \mathbb{T}_{u\downarrow})$ and v some element in H, we define

$$u_*\, M(v) = M(u \wedge v)$$

where $M(u \wedge v)$ is equipped with the structure of a $\mathbb{T}(v)$ algebra via the morphism of theories $\mathbb{T}(v) \to \mathbb{T}(u \wedge v)$: any operation at the level v is sended to an operation at the level $u \wedge v$ and thus acts on $M(u \wedge v)$. If $w \leqslant v$ in H, the restriction morphism $u_*\, M(v) \to u_*\, M(w)$ is simply $M(u \wedge w \leqslant u \wedge v)$. If $v = \bigvee_{i \in I} v_i$ in H, $u \wedge v = \bigvee_{i \in I} (u \wedge v)$ in H and in $u{\downarrow}$; so $u_*\, M$ is a sheaf because M is one. Finally if $m : M \to N$ is a morphism in $\mathrm{Sh}(u{\downarrow}, \mathbb{T}_{u\downarrow})$, we define $u_*\, m$ by

$$(u_*\, m)_v = m_{u \wedge v}.$$

This is clearly a morphism in Sh(H, T) and this completes the description of u_*.

We define, for any A in Sh(H, T), a morphism

$$\alpha_A : A \to u_*\, u^*\, A;$$

for any v in H, $\alpha_A(v) : A(v) \to A(u \wedge v)$ is the restriction morphism $A(u \wedge v \leqslant v)$. Now if M is some object in $\mathrm{Sh}(u{\downarrow}, \mathbb{T}_{u\downarrow})$ and $f : A \to u_*\, M$ some morphism in Sh(H, T), we define $m : u^*\, A \to M$ in $\mathrm{Sh}(u{\downarrow}, \mathbb{T}_{u\downarrow})$ by $m_v = f_v$

$$m_v : u^*\, A(v) = A(v) \to u_*\, M(v) = M(v)$$

for any $v \leqslant u$. Clearly, m is the unique morphism from $u^*\, A$ to M such that $u^*\, m \circ \alpha_A = f$. So u_* is right adjoint to u^*.

u_* is faithful. Indeed, consider two different morphisms $m, n : M \to N$ in $\mathrm{Sh}(u{\downarrow}, \mathbb{T}_{u\downarrow})$. There is some $v \leqslant u$ in H such that $m_v \neq n_v$. But then $(u_*\, m)_v = m_v \neq n_v = (u_*\, n)_v$. So $u_*\, m \neq u_*\, n$ and u_* is faithful.

If M is some object in $\mathrm{Sh}(u{\downarrow}, \mathbb{T}_{u\downarrow})$ and v some element in $u{\downarrow}$, one has $u^*\, u_*\, M(v) = M(v)$; therefore $u^*\, u_*\, M \cong M$. Now if M, N are objects in $\mathrm{Sh}(u{\downarrow}, \mathbb{T}_{u\downarrow})$ and $f : u_*\, M \to u_*\, N$ is some arrow in Sh(H, T), we shall prove that $f = u_*\, u^*\, f$ and this will show that u_* is full. From $u^*\, u_*\, M \cong M$ we deduce that $u^*\, f$ has the form $u^*\, f : M \to N$. Now for any v in H, $(u_*\, u^*\, f)_v = (u^*\, f)_{u \wedge v} = f_{u \wedge v}$. But $(u_*\, M)(u \wedge v) = (u_*\, M)(v)$ and $(u_*\, M)(u \wedge v \leqslant w)$ is the identity morphism on

$M(u \wedge v)$; so $f_{u \wedge v} = f_v$ and finally $u_* u^* f = f$. Thus u_* is full.

We now turn our attention to the definition of $u_!$. If M is some object in $Sh(u\downarrow, \mathbb{T}_{u\downarrow})$, $u_!M$ will be the sheaf associated to some presheaf u'M that we define now. For any v in $\mathbb{H}$

$$(u'M)(v) = \begin{cases} M(v) & \text{if} \quad v \leqslant u \\ \mathcal{O}_0(v) & \text{if} \quad v \not\leqslant u \end{cases}$$

where $\mathcal{O}_0(v)$ denotes the zero-operations (the constants) of the theory $\mathbb{T}(v)$. If $w \leqslant v$ in $\mathbb{H}$, we have a restriction morphism $(u'M)(w \leqslant v)$ which is $M(w \leqslant v)$ if $v \leqslant u$ and the composite

$$\mathcal{O}_0(v) \xrightarrow{\mathcal{O}_0(w \leqslant v)} \mathcal{O}_0(w) \longrightarrow (u'M)(w)$$

if $v \not\leqslant u$. (Recall that $\mathcal{O}_0(w)$ is the initial object in $\underline{Sets}^{\mathbb{T}(w)}$). u'M is an object in $Pr(\mathbb{H}, \mathbb{T})$ and $u_!M$ is its associated sheaf. If $m : M \to N$ is a morphism in $Sh(u\downarrow, \mathbb{T}_{u\downarrow})$ and v some element in $\mathbb{H}$, we define

$$(u'm)_v = \begin{cases} m_v & \text{if} \quad v \leqslant u \\ id_{\mathcal{O}_0(v)} & \text{if} \quad v \not\leqslant u. \end{cases}$$

u'm is a morphism in $Pr(\mathbb{H}, \mathbb{T})$ and applying the associated sheaf functor, we get the morphism $u_!m$ in $Sh(\mathbb{H}, \mathbb{T})$. This completes the description of $u_!$.

If A is some object in $Sh(\mathbb{H}, \mathbb{T})$, we define a morphism $\beta'_A : u' u^* A \to A$ in $Pr(\mathbb{H}, \mathbb{T})$ by

$$\begin{cases} \beta'_A(v) : u' u^* A(v) = A(v) \to A(v) \text{ is the identity if } v \leqslant u \\ \beta'_A(v) : u' u^* A(v) = F(0) \to A(v) \text{ is trivial if } v \not\leqslant u. \end{cases}$$

(We recall that F(0) is an initial object). Applying the associated sheaf functor to β'_A we get a morphism $\beta_A : u_! u^* A \to A$ in $Sh(\mathbb{H}, \mathbb{T})$. We need to prove its universal property.

Consider M in $Sh(u\downarrow, \mathbb{T}_{u\downarrow})$ and $f : u_!M \to A$ in $Sh(\mathbb{H}, \mathbb{T})$. Composing with the universal morphism $u'M \to u_!M$ we get a morphism $\overline{f} : u'M \to A$ in $Pr(\mathbb{H}, \mathbb{T})$. For any v in $u\downarrow$, whe choose $g(v) : M(v) = u'M(v) \to A(v) = u^* A(v)$ to be $\overline{f}(v)$. This defines a morphism $g : M \to u^* A$ in $Sh(u\downarrow, \mathbb{T}_{u\downarrow})$ such that $\beta_A \circ u_!(g) = f$.

This also proves the uniqueness of g. So $u_!$ is left adjoint to u^* and, because u_* is full and faithful, the same holds for $u_!$ (by [21] - 16 - 8 - 9).

If $m : M \rightarrowtail N$ is a monomorphism in $Sh(u{\downarrow}, \mathbb{T}_{u\downarrow})$, each $m(v) : M(v) \to N(v)$ is injective, for any v in $u{\downarrow}$. So $u'm(v)$ is equal to the injection $m(v)$ when $v \leqslant u$ and to the identity on $F(0)$ when $v \nleqslant u$. In any case, $u'm(v)$ is injective, so $u'm$ is a monomorphism in $Pr(\mathbb{H}, \mathbb{T})$. By exactness of the associated sheaf functor, $u_!(m)$ is a monomorphism in $Sh(\mathbb{H}, \mathbb{T})$. This proves that $u_!$ preserves monomorphisms.

To prove that $u_!$ creates monomorphisms, consider an object M in $Sh(u{\downarrow}, \mathbb{T}_{u\downarrow})$ and a monomorphism $f : A \rightarrowtail u_!M$ in $Sh(\mathbb{H}, \mathbb{T})$, we shall prove that $f \cong u_!\ u^*(f)$ which concludes the proof, for $u^* f$ is a monomorphism in $Sh(u{\downarrow}, \mathbb{T}_{u\downarrow})$. ($u^*$ preserves limits).

u^* is a reflection of the full and faithful functor $u_!$; therefore $u^*\ u_!(M) \cong M$. (Cfr. [21] - 16 - 5 - 4). u^* has a left adjoint $u_!$; thus u^* preserves monomorphisms and $u^*(f) : u^*(A) \to u^*\ u_!(M) \cong M$ is a monomorphism in $Sh(u{\downarrow}, \mathbb{T}_{u\downarrow})$. It remains to prove that $u_!\ u^*\ f \cong f$. Consider the following diagram in $Pr(\mathbb{H}, \mathbb{T})$:

$$\begin{array}{ccc} u'\ u^* A & \longrightarrow & A \\ {\scriptstyle u'\ u^*(f)} \downarrow & & \downarrow {\scriptstyle f} \\ u'M & \longrightarrow & u_!M \end{array}$$

where the horizontal mappings arise from the adjunctions. If $v \leqslant u$, the v-component of this diagram is

$$\begin{array}{ccc} A(v) & \xrightarrow{\ id\ } & A(v) \\ {\scriptstyle f(v)} \downarrow & & \downarrow {\scriptstyle f(v)} \\ M(v) & \xrightarrow[\ id\ ]{} & M(v) \end{array}$$

which is a pullback. If $v \nleqslant u$, the v-component of the same diagram is

$$\begin{array}{ccc} F(0) & \longrightarrow & A(v) \\ \| & & \downarrow {\scriptstyle f(v)} \\ F(0) & \longrightarrow & u_!M(v) \end{array}$$

which is again a pullback because the image of the mapping $F(0) \to u_!M(v)$ is the smallest subobject of $u_!M(v)$ - (the set of constants) - and is thus contained in A(v). Therefore our diagram in Pr(H, T) is a pullback; applying the exact associated sheaf functor, we obtain a pullback in Sh(H, T) :

$$\begin{array}{ccc} u_! \, u^* A & \longrightarrow & A \\ {\scriptstyle u_! u^* f}\downarrow & & \downarrow{\scriptstyle f} \\ u_! M & \xrightarrow[\text{id}]{} & u_! M \end{array}$$

The lower horizontal morphism is an isomorphism. So the upper one is an isomorphism. This gives us the isomorphism $f \cong u_! \, u^* f$.

Now consider the image $u_! \, u^* A \twoheadrightarrow I \rightarrowtail A$ in Sh(H, T) of the canonical morphism $u_! \, u^* A \to A$. This image is obtained from the image $u' \, u^* A \twoheadrightarrow J \rightarrowtail A$ in Pr(H, T) by applying the associated sheaf functor. The exactness of the associated sheaf functor implies that it takes a Heyting subobject on a Heyting subobject. So we have to show that J is a Heyting subobject of A in Pr(H, T). But in Pr(H, T), all the notions which appear in the definition of a Heyting subobject are computed pointwise : so it suffices to prove that each J(v) is a Heyting subobject of A(v). But if $v \leqslant u$, J(v) is A(v) and if $v \not\leqslant u$, J(v) is the set of constants in A(v), i.e. the smallest subobject of A(v); in both cases, J(v) is obviously a Heyting subobject of A(v).

Finally suppose that $0 \to A$ is a monomorphism in Sh(H, T). This means that for each v in H, A(v) contains (injectively) the set F(0) of constants of the theory T(v). Now look at the canonical morphism $\beta'_A : u' \, u^* A \to A$ in Pr(H, T); if $v \leqslant u$, $\beta'_A(v)$ is the identity on A(v) and if $v \not\leqslant u$, $\beta'_A(v)$ is the inclusion of F(0) in A(v) : thus β'_A is a monomorphism and the same holds for β_A, by exactness of the associated sheaf functor.

Now if $A = a \, F \, h_1$ (we know that $0 \to a \, F \, h_1$ is a monomorphism; proposition I - 7) and $v \in H$

$$u' \, u^*(a \, F \, h_1)(v) = \begin{cases} a \, F \, h_1(v) = a \, F \, h_u(v) & \text{if} \quad v \leqslant u \\ F \, 0 = F \, h_u(v) & \text{if} \quad v \not\leqslant u. \end{cases}$$

This implies that $u_! \, u^*(a \, F \, h_1) = a \, F \, h_u$. ■

§ 3. THE FRAME OF FORMAL INITIAL SEGMENTS

In § 2 we have shown some relations between the categories $\mathrm{Sh}(u\!\downarrow, \mathbb{T}_{u\downarrow})$ and $\mathrm{Sh}(\mathbb{H}, \mathbb{T})$ for any u in H. In this paragraph, we consider the full imbeddings $\mathcal{U} \hookrightarrow \mathrm{Sh}(\mathbb{H}, \mathbb{T})$ which satisfy essentially the same properties as $u_!$ in theorem 5; we call them "formal initial segments" of $\mathrm{Sh}(\mathbb{H}, \mathbb{T})$. These formal initial segments are meant to "copy" the elements of H. So it is not all that surprising that in turn they give rise to a frame. But the proof of this fact is rather arduous. In the next paragraph we will show that this new frame contains H as a subframe.

The observant reader will note some minor differences between the conditions of theorem 5 and definition 6 (completed by the axiomatic consequences described in proposition 7). We are unable to explain these differences; we simply point out that they vanish when the morphism $0 \to A$ is always a monomorphism in $\mathrm{Sh}(\mathbb{H}, \mathbb{T})$: in particular this is the case when $\mathbb{T}$ has, at each level, a single generic constant or none at all. The notion of formal initial segment is essentially a tool which makes possible the proof of the characterization theorem in chapter 4, so we may freely adapt it to make the proofs work : this is what we do. However, in chapter 6, the notion of formal initial segment itself turns out to be interesting, but then the theory $\mathbb{T}$ considered there is the theory of modules on a ring R and thus $\mathbb{T}$ has a single constant : therefore the differences between theorem 5 and definition 6 vanish.

Definition 6.

Let C be a category equivalent to some category $\mathrm{Sh}(\mathbb{H}, \mathbb{T})$. *A formal initial segment in C is a full subcategory* $\mathcal{U}$ *of C such that, if* $u_! : \mathcal{U} \hookrightarrow C$ *is the canonical inclusion*

(F 1) $u_!$ *has a right adjoint* u^*

(F 2) u^* *has a right adjoint* u_*

(F 3) *if* $0 \to M$ *is a monomorphism in C with codomain M in* $\mathcal{U}$, $u_!$ *creates monomorphisms with codomain M*

(F 4) *if* $0 \to A$ *is a monomorphism in C, the canonical morphism* $u_! u^* A \to A$ *arising from the adjunction* $u_! \dashv u^*$ *is a monomorphism*

(F 5) *if* $0 \to A$ *is a monomorphism in C, the canonical monomorphism* $u_! u^* A \to A$ *is a Heyting subobject.*

For the sake of brevity, we shall often use the notation uA or u(A) for

$u_! u^* A$.

<u>Proposition 7.</u>

Let C *be a category equivalent to some category* $\mathrm{Sh}(\mathbb{H}, \mathbb{T})$. *Let* $\mathcal{U}$ *be a formal initial segment in* C.

The following conditions hold :

(F 6) u_* *is full and faithful.*

(F 7) $u_!$ *reflects monomorphisms.*

(F 8) *if* $0 \to M$ *is a monomorphism in* $\mathcal{U}$, $u_!$ *preserves monomorphisms with codomain* M.

As $u_!$ is full and faithful so is u_* (cfr. [21] - 16 - 8 - 9). Also $u_!$ reflects monomorphisms because it is faithful. Now consider a monomorphism $m : N \to M$ in $\mathcal{U}$ with $0 \to M$ a monomorphism in $\mathcal{U}$. Consider the image of m in C

$$\begin{array}{ccc} N & \xrightarrow{\quad m \quad} & M \\ & {\scriptstyle p}\searrow \quad \nearrow{\scriptstyle i} & \\ & I & \end{array} .$$

By (F 3), I is (up to an isomorphism) in $\mathcal{U}$; thus the whole diagram is in $\mathcal{U}$ and is invariant under u^* (cfr. [21] - 16 - 5 - 4). But u^* has a right and a left adjoint; so it preserves monomorphisms and regular epimorphisms. This shows that I still is the image of m in $\mathcal{U}$. But m is a monomorphism in $\mathcal{U}$; so p is both a monomorphism and a regular epimorphism in $\mathcal{U}$: it is an isomorphism. So m is isomorphic to i, a monomorphism in C. ■

<u>Proposition 8.</u>

Let C *be a category equivalent to some category* $\mathrm{Sh}(\mathbb{H}, \mathbb{T})$. *Let* $\mathcal{U}$ *be a formal initial segment in* C. *Let* $0 \rightarrowtail A$ *and* $i : B \rightarrowtail A$ *be monomorphisms in* C. *The following square is a pullback in* C

$$\begin{array}{ccc} u(B) & \xrightarrow{\ \beta_B\ } & B \\ {\scriptstyle u(i)}\downarrow & & \downarrow{\scriptstyle i} \\ u(A) & \xrightarrow[\ \beta_A\ ]{} & A \end{array}$$

where the horizontal arrows are those arising from the adjunction $u_! \dashv u^*$. *In other words*

$$u(B) = B \cap u(A).$$

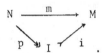

Consider the following diagram, where the square is a pullback in C

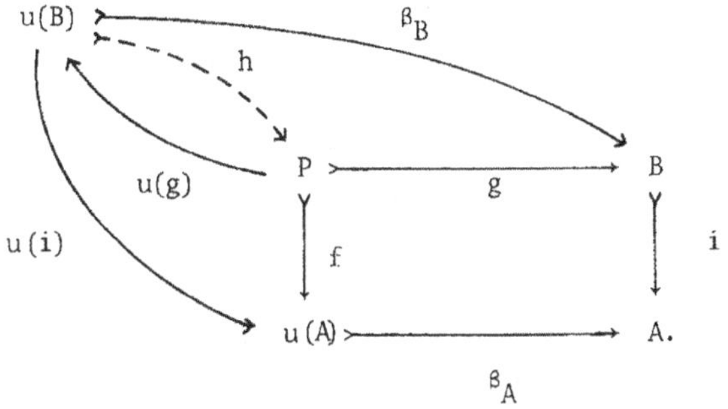

By F 3, P is an object in U and so $u(P) \cong P$ ([21] - 16 - 5 - 4). So $u(g)$ is a morphism from P to $u(B)$. Now β_A and β_B are the canonical morphisms arising from the adjunction $u_! \dashv u^*$; so the outside diagram is commutative and we get the factorization h through the pullback. As $0 \to A$ is a monomorphisms so is $0 \to B$. Therefore β_B is a monomorphism (F 4) and so is h. Now compute

$$\begin{aligned} f \circ h \circ u(g) &= u(i) \circ u(g) \\ &= u(i\ g) \\ &= u(\beta_A\ f) \\ &= u(\beta_A) \circ u(f) \\ &= f \end{aligned}$$

(by [21] 16 - 5 - 5, 16 - 5 - 4 and the fact that f is in U)

$$\begin{aligned} g \circ h \circ u(g) &= \beta_B \circ u(g) \\ &= g \circ \beta_P \\ &= g \end{aligned}$$

(by [21] - 16 - 5 - 4 and the fact that P is in U). So the monomorphism h has a section $u_!\ u^*(g)$: it is an isomorphism. This proves the proposition. ■

<u>Proposition 9.</u>

Let C be a category equivalent to some category $\mathrm{Sh}(\mathbb{H}, \mathbb{T})$. *Let U and V be two formal initial segments in C with $U \leqslant V$.*

Then U is a formal initial segment in V.

Clearly, this last statement makes sense even if V is not equivalent to some category of algebraic sheaves. We denote by $w_! : U \to V$ the canonical inclusion. Clearly $u_! = v_! \circ w_!$ and so $w_! = v^* \circ v_! \circ w_! = v^* \circ u_!$.

We define $w^* : V \to U$ by $w^* = u^* \circ v_!$. w^* is right adjoint to $w_!$ because of the natural bijections, for M in U and N in V

$$\begin{aligned}(M, w^* N) &\cong (M, u^* v_! N)\\ &\cong (u_! M, v_! N)\\ &\cong (M, N)\\ &\cong (w_! M, N).\end{aligned}$$

From $u_! = v_! \circ w_!$ we deduce $u^* = w^* \circ v^*$.

We define $w_* : U \to V$ by $w_* = v^* u_*$. w_* is right adjoint to w^* because of the natural bijections, for M in U and N in V

$$\begin{aligned}(N, w_* M) &\cong (N, v^* u_* M)\\ &\cong (v_! N, u_* M)\\ &\cong (u^* v_! N, M)\\ &\cong (w^* N, M).\end{aligned}$$

From $u^* = w^* \circ v^*$ we deduce $u_* = v_* \circ w_*$.

If $\alpha^U : id_C \Rightarrow u_* u^*$, $\alpha^V : id_C \Rightarrow v_* v^*$, $\beta^U : u_! u^* \Rightarrow id_C$, $\beta^V : v_! v^* \Rightarrow id_C$ are the canonical natural transformations defining the adjunctions, then the canonical transformation $\alpha^W : id_V \Rightarrow w_* w^*$ and $\beta^W : w_! w^* \Rightarrow id_V$ are given by the following morphisms, for any object N in V

$$\alpha^W_N : N \cong v^* N \xrightarrow{v^*(\alpha^U_N)} v^* u_* u^* N = w_* w^* N$$

$$\beta^W_N : w_! w^* N \cong u_! u^* N \xrightarrow{\beta^U_N} v_! N = N$$

as follows from the bijections just described.

If $0 \to N$ is a monomorphism in V, $0 \to N$ is a monomorphism in C (by F 8) and β^U_N is a monomorphism and a Heyting subobject in C (by F 4 - F 5) and thus also in N because $v_!$ preserves unions and intersections of subobjects of N (by F 1 - F 8). Finally if M is in U and $0 \to M$, $N \to M$ are monomorphisms in V, they are also monomorphisms in C (by F 8) and thus N is in U (by F 3). This concludes the proof that U is an initial segment in V.

<u>Corollary 10.</u>

For $\mathcal{U}$ and $\mathcal{V}$ and their various functors given as in proposition 9, *the following equalities hold :*

$$u_! \, u^* \, v_! \, v^* = u_! \, u^* = v_! \, v^* \, u_! \, u^*$$

$$u_* \, u^* \, v_* \, v^* = u_* \, u^* = v_* \, v^* \, u_* \, u^*$$

$$v_* \, v^*(\alpha^{\mathcal{U}}_A) = \alpha^{\mathcal{U}}_{v_* \, v^* A} : v_* \, v^* A \to u_* \, u^* A$$

$$v_! \, v^*(\beta^{\mathcal{U}}_A) = \beta^{\mathcal{U}}_{v_! \, v^* A} : u_! \, u^* A \to v_! \, v^* A.$$

The following equalities hold

$$u_! \, u^* \, v_! \, v^* = u_! \, w^* \, v^* = u_! \, u^*$$

$$v_* \, v^* \, u_* \, u^* = v_* \, w_* \, u^* = u_* \, u^*$$

and so for any object A in C we have natural morphisms

$$v_* \, v^*(\alpha^{\mathcal{U}}_A) : v_* \, v^* A \to u_* \, u^* A$$

$$\beta^{\mathcal{U}}_{v_! \, v^* A} : u_! \, u^* A \to v_! \, v^* A.$$

But we also have

$$v_! \, v^* \, u_! \, u^* = v_! \, w_! \, u^* = u_! \, u^*.$$

By taking their right adjoints we get

$$u_* \, u^* \, v_* \, v^* = u_* \, w^* \, v^* = u_* \, u^*.$$

So for any A in C we have natural morphisms

$$\alpha^{\mathcal{U}}_{v_* \, v^* A} : v_* \, v^* A \to u_* \, u^* A$$

$$v_! \, v^*(\beta^{\mathcal{U}}_A) : u_! \, u^* A \to v_! \, v^* A.$$

In fact the following equalities hold

$$v_* \, v^*(\alpha^{\mathcal{U}}_A) = \alpha^{\mathcal{U}}_{v_* \, v^* A}$$

$$v_! \, v^*(\beta^{\mathcal{U}}_A) = \beta^{\mathcal{U}}_{v_! \, v^* A}.$$

Indeed, the following triangular equality arises from the adjunction $v^* \dashv v_*$

$$\begin{array}{ccc} v^* & \xrightarrow{v^* \alpha^{\mathcal{V}}} & v^* v_* v^* \\ & \diagdown\!\!\diagdown & \| \\ & & v^*. \end{array}$$

Now $v_! v^*$ is left adjoint to $v_* v^*$; thus $v_* v^*(\alpha_A^{\mathcal{U}})$ is the unique morphism from $v_* v^* A$ to $u_* u^* A$ such that

$$\alpha_A^{\mathcal{U}} \circ \beta_A^{\mathcal{V}} = \beta^{\mathcal{V}}_{u_* u^* A} \circ v_! v^* v_* v^*(\alpha_A^{\mathcal{U}}).$$

But the following equalities hold :

$$\beta^{\mathcal{V}}_{u_* u^* A} \circ v_! v^*(\alpha^{\mathcal{U}}_{v_* v^* A})$$

$$= \alpha^{\mathcal{U}}_{v_* v^* A} \circ \beta^{\mathcal{V}}_{v_* v^* A} \qquad \text{naturality}$$

$$= \alpha^{\mathcal{U}}_{v_* v^* A} \circ \beta^{\mathcal{V}}_{v_* v^* A} \circ v_! v^*(\alpha_A^{\mathcal{V}})$$

$$= \alpha^{\mathcal{U}}_{v_* v^* A} \circ \alpha_A^{\mathcal{V}} \circ \beta_A^{\mathcal{V}} \qquad \text{naturality}$$

$$= u_* u^* \alpha_A^{\mathcal{V}} \circ \alpha_A^{\mathcal{U}} \circ \beta_A^{\mathcal{V}} \qquad \text{naturality}$$

$$= u_* w^* v^*(\alpha_A^{\mathcal{V}}) \circ \alpha_A^{\mathcal{U}} \circ \beta_A^{\mathcal{V}}$$

$$= \alpha_A^{\mathcal{U}} \circ \beta_A^{\mathcal{V}}$$

and they show that $v_* v^* \alpha_A^{\mathcal{U}} = \alpha^{\mathcal{U}}_{v_* v^* A}$.

In the same way the triangular equality

$$\begin{array}{ccc} v^* & = & v^* v_! v^* \\ & \diagdown\!\!\diagdown & \downarrow {\scriptstyle v^* \beta^{\mathcal{V}}} \\ & & v^* \end{array}$$

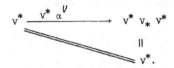

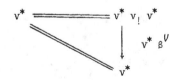

arises from the adjunction $v_! \dashv v^*$. Now $v_* v^*$ is right adjoint to $v_! v^*$; so $v_! v^*(\beta^{U}_{A})$ is the unique morphism from $u_! u^*(A)$ to $v_! v^*(A)$ such that

$$v_* v^* v_! v^*(\beta^{U}_{A}) \circ \alpha^{V}_{u_! u^* A} = \alpha^{V}_{A} \circ \beta^{U}_{A}.$$

But the following equalities hold

$$v_* v^* \beta^{U}_{v_! v^* A} \circ \alpha^{V}_{u_! u^* A}$$

$$= \alpha^{V}_{v_! v^* A} \circ \beta^{U}_{v_! v^* A} \qquad \text{naturality}$$

$$= v_* v^* \beta^{V}_{A} \circ \alpha^{V}_{v_! v^* A} \circ \beta^{U}_{v_! v^* A}$$

$$= \alpha^{V}_{A} \circ \beta^{V}_{A} \circ \beta^{U}_{v_! v^* A} \qquad \text{naturality}$$

$$= \alpha^{V}_{A} \circ \beta^{U}_{A} \circ u_! u^* \beta^{V}_{A} \qquad \text{naturality}$$

$$= \alpha^{V}_{A} \circ \beta^{U}_{A} \circ u_! w^* v^* \beta^{V}_{A}$$

$$= \alpha^{V}_{A} \circ \beta^{U}_{A}$$

and they show that $\beta^{U}_{v_! v^* A} = v_! v^*(\beta^{U}_{A})$. ■

We prove now our theorem II - 11 which is one of the crucial points of this work. The characterization theorem of chapter 4 and the representation theorems of chapters 5 - 6 - 7 - 8 are based on it.

Theorem 11.

Let C *be a category equivalent to some category* Sh($\mathbb{H}$, $\mathbb{T}$). *The formal initial segments of* C *(considered up to an equivalence) form a frame for the ordering given by the usual inclusion of subcategories of* C.

Let H denote the class ordered by inclusion, of equivalence classes of formal initial segments in C. To avoid the consideration of equivalence classes, we choose canonically a formal initial segment in each equivalence class. This is the one saturated for isomorphisms.

$\mathcal{H}$ has a smallest element $\mathcal{O}$ which consists of the initial object(s) of $\mathcal{C}$; o^* is thus the constant functor on 0 and o_* is completely defined by $o_*(0) = 1$. It is clear that, for any object A in Sh($\mathbb{H}$, $\mathbb{T}$) one has

$$(o_!(0), A) \cong (0, A) \cong 1 \cong (0, o^* A)$$
$$(o^* A, 0) \cong (0, 0) \cong 1 \cong (A, 1) \cong (A, o^*(0))$$

and so we obtain the adjunctions

$$o_! \dashv o^* \dashv o^*.$$

Now if $0 \to A$ is a monomorphism in $\mathcal{C}$, thus the canonical morphism $o_!\, o^*(A) = 0 \to A$ is a monomorphism and it is a Heyting subobject because 0 is the smallest subobject of A. Now if M is in $\mathcal{O}$, i.e. if $M = 0$, any monomorphism $f : A \rightarrowtail 0$ has a section $0 \to A$ and hence is an isomorphism; so A is in $\mathcal{O}$.

$\mathcal{H}$ has a greatest element I which is the whole category $\mathcal{C}$, the functors $i_!$ i^* and i_* are the identity on $\mathcal{C}$. Conditions F 1 - F 2 - F 3 - F 4 obviously are satisfied and so is condition F 5 since an object is always a Heyting subobject of itself.

We consider now two formal initial segments $\mathcal{U}$ and $\mathcal{V}$ of $\mathcal{C}$ and we construct their infimum $\mathcal{U} \wedge \mathcal{V}$ in $\mathcal{H}$. First of all, the subcategory $\mathcal{U} \wedge \mathcal{V}$ is just the intersection $\mathcal{U} \cap \mathcal{V}$ of the two given subcategories; $(u \wedge v)_!$ is the canonical inclusion of $\mathcal{U} \cap \mathcal{V}$ in $\mathcal{C}$. For the sake of brevity, we write $\mathcal{W} = \mathcal{U} \wedge \mathcal{V}$, $w_! = (u \wedge v)_!$, and so on ...

To prove the existence of w^* and w_*, we prove first that $u_!\, u^*\, v_!\, v^*$ is a functor naturally isomorphic to $v_!\, v^*\, u_!\, u^*$. But $u_!$, u^*, $v_!$ and v^* have right adjoints : so $u_!\, u^*\, v_!\, v^*$ and $v_!\, v^*\, u_!\, u^*$ are cocontinuous functors; they will be isomorphic as soon as they are isomorphic on a dense subcategory of $\mathcal{C}$ (cfr. [21] - 17 - 2 - 7). By proposition I - 7, it suffices to prove that both functors coincide on those objects A such that the morphism $0 \to A$ is a monomorphism. By F 4 and proposition 8, if $0 \to A$ is a monomorphism, then $u_!\, u^*\, v_!\, v^*(A)$ and $v_!\, v^*\, u_!\, u^*(A)$ coincide with the intersection of the two subobjects $\beta_A^{\mathcal{U}} : u(A) \to A$ and $\beta_A^{\mathcal{V}} : v(A) \to A$. In particular this double application of proposition 8 and the uniqueness of the pullback show that $u(\beta_A^{\mathcal{V}}) = \beta_{uA}^{\mathcal{V}}$ and $v(\beta_A^{\mathcal{U}}) = \beta_{vA}^{\mathcal{U}}$.

$$\begin{array}{ccc} uA \cap vA & \xrightarrow{v(\beta_A^{\mathcal{U}}) \,=\, \beta_{vA}^{\mathcal{U}}} & vA \\ {\scriptstyle u(\beta_A^{\mathcal{V}}) \,=\, \beta_{uA}^{\mathcal{V}}}\big\downarrow & & \big\downarrow{\scriptstyle \beta_A^{\mathcal{V}}} \\ uA & \xrightarrow[\beta_A^{\mathcal{U}}]{} & A \end{array}$$

By naturality and uniqueness of the factorization through a pullback, this implies that the functors $u_! \, u^* \, v_! \, v^*$ and $v_! \, v^* \, u_! \, u^*$ coincide also on any morphism $f : A \to B$ in C, where $0 \to A$ and $0 \to B$ are monomorphisms. Finally we have shown that

$$u_! \, u^* \, v_! \, v^* \cong u_! \, u^* \cap v_! \, v^* \cong v_! \, v^* \, u_! \, u^*,$$

but this formula also shows that this functor takes its values in $\mathcal{U} \cap \mathcal{V}$; we define it to be w^*.

We have just shown that, when $0 \to A$ is a monomorphism,

$$\begin{aligned} \beta_A^{\mathcal{U}} \circ \beta_{uA}^{\mathcal{V}} &= \beta_A^{\mathcal{U}} \circ u(\beta_A^{\mathcal{V}}) \\ &= \beta_A^{\mathcal{V}} \circ \beta_{vA}^{\mathcal{U}} \\ &= \beta_A^{\mathcal{V}} \circ v(\beta_A^{\mathcal{U}}). \end{aligned}$$

Again by [21] - 12 - 2 - 7 and proposition I - 1, this implies an equality between natural transformations

$$\begin{aligned} \beta^{\mathcal{U}} \circ (\beta^{\mathcal{V}} * u) &= \beta^{\mathcal{U}} \circ (u * \beta^{\mathcal{V}}) \\ &= \beta^{\mathcal{V}} \circ (\beta^{\mathcal{U}} * v) \\ &= \beta^{\mathcal{V}} \circ (v * \beta^{\mathcal{U}}); \end{aligned}$$

we choose this natural transformation $w_! \, w^* \Rightarrow \mathrm{id}$ to be $\beta^{\mathcal{W}}$. $\beta^{\mathcal{W}}$ has the required universal property. Indeed, let M be some object in $\mathcal{U} \cap \mathcal{V}$, A any object in C and $f : M \to A$ any morphism in C. M is in $\mathcal{U}$ and so we obtain a unique morphism $g : M \to u^* A$ in $\mathcal{U}$ such that $\beta_A^{\mathcal{U}} \circ g = f$ in C. Now M is in $\mathcal{V}$ and the morphism $g : M \to u_! \, u^* A$ factors uniquely through $\beta_{u_! \, u^* A}^{\mathcal{V}}$ by a morphism $h : M \to v^* \, u_! \, u^* A$

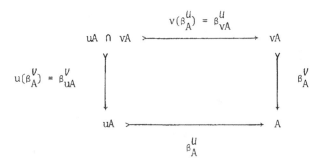

in $\mathcal{V}$. So we get a unique factorization $h : M \to w^* A$ of f through $\beta_A^{\mathcal{W}}$; the domain and the codomain of h are in the full subcategory $\mathcal{U} \cap \mathcal{V}$, so h is in $\mathcal{U} \cap \mathcal{V}$. Thus w^* is right adjoint to $w_!$.

The construction of $\beta^{\mathcal{W}}$ shows that $\beta_A^{\mathcal{W}}$ is a monomorphism as soon as $0 \to A$ is a monomorphism. Now the intersection of two Heyting subobjects is a Heyting subobject (theorem 4). So if $0 \to A$ is a monomorphism, $\beta_A^{\mathcal{W}}$ is a Heyting subobject as the intersection of the Heyting subobjects $\beta_A^{\mathcal{U}}$ and $\beta_A^{\mathcal{V}}$. Moreover, if $0 \to M$ and $A \to M$ are monomorphisms in C with M in $\mathcal{U} \cap \mathcal{V}$, then by F 3, A is in $\mathcal{U}$ and A is in $\mathcal{V}$; so A is in $\mathcal{U} \cap \mathcal{V}$. This shows that $w_!$ and w^* satisfy F 3 - F 4 - F 5.

We still need to construct w_* right adjoint to w^*. But from the isomorphism

$$u_! \, u^* \, v_! \, v^* \cong v_! \, v^* \, u_! \, u^*$$

we obtain, by taking the right adjoints of both sides,

$$v_* \, v^* \, u_* \, u^* \cong u_* \, u^* \, v_* \, v^*.$$

Now we define

$$w_* = v_* \, v^* \, u_* \, u^* \, w_! \cong u_* \, u^* \, v_* \, v^* \, w_!$$

which is a functor from $\mathcal{U} \cap \mathcal{V}$ to C.
By the adjunction just mentioned, we have natural isomorphisms

$$(u_! \, u^* \, v_! \, v^* A, B) \cong (A, v_* \, v^* \, u_* \, u^* B)$$

with A and B objects in C.
In particular for A in C and M in $\mathcal{U} \cap \mathcal{V}$ one has natural isomorphisms

$$\begin{aligned}(w^* A, M) &\cong (u_! \, u^* \, v_! \, v^* A, w_! M) \\ &\cong (A, v_* \, v^* \, u_* \, u^* \, w_! M) \\ &\cong (A, w_* M)\end{aligned}$$

which show that w_* is right adjoint to w^*. This concludes the proof of the fact that $\mathcal{W} = \mathcal{U} \wedge \mathcal{V}$ is an initial segment of C. Its definition as the intersection of the categories $\mathcal{U}$ and $\mathcal{V}$ immediately shows that $\mathcal{W}$ is the infimum of $\mathcal{U}$ and $\mathcal{V}$ in the ordered class of initial segments of C.

We shall now construct the supremum of a family $(\mathcal{U}_i)_{i \in I}$ of formal initial segments of C, for a non empty indexing set I. We already know that for any non empty finite subset J of I, the infimum $\bigwedge_{i \in I} \mathcal{U}_i$ exists. But clearly $\bigvee_{i \in I} \mathcal{U}_i$ exists if and only if

$$\bigvee_J \Big(\bigwedge_{i \in J} \mathcal{U}_i \Big), \text{ where } J \subseteq I;\ J \neq \phi;\ J \text{ finite},$$

exists and in this case, they are equal. Now the second supremum is indexed by a filtered family of J's. This allows us to reduce the problem to the case of a filtered family $(\mathcal{U}_i)_{i\in I}$ of formal initial segments.

So let $(\mathcal{U}_i)_{i\in I}$ be a filtered family of formal initial segments of C. If $\mathcal{U}_i \leqslant \mathcal{U}_j$ then, by applying corollary 10, we obtain for every object A in C a morphism

$$u_j(\beta_A^{\mathcal{U}_i}) = \beta_{u_j(A)}^{\mathcal{U}_i} : u_i(A) \rightarrow u_j(A).$$

Consider the diagram whose vertices are all the $u_i(A)$ and whose arrows are those morphisms which we just described. Define $u(A)$ to be the (filtered) colimit of this diagram

$$u(A) = \varinjlim_{i\in I} u_i(A).$$

If $f : A \rightarrow B$ is any morphism in C, the naturality of the $\beta^{\mathcal{U}_i}$ implies that the morphisms

$$u_i(f) : u_i(A) \rightarrow u_i(B)$$

induce a natural transformation between the diagrams defining $u(A)$ and $u(B)$. As a consequence, we get a factorization $u(f) : u(A) \rightarrow u(B)$ and finally u becomes an endofunctor of C. Define $\mathcal{U} = \bigvee_{i\in I} \mathcal{U}_i$ to be the full image of u, saturated under isomorphisms. $u_!$ is the inclusion of $\mathcal{U}$ in C and u^* is the corestriction of u to $\mathcal{U}$. Thus $u = u_! u^*$.

If A is any object in C, we have a cone $(\beta_A^{\mathcal{U}_i} : u_i(A) \rightarrow A)_{i\in I}$ on the diagram defining $u(A)$; indeed if $\mathcal{U}_i \leqslant \mathcal{U}_j$, the pullback we constructed in order to define $\mathcal{U}_i \wedge \mathcal{U}_j$ reduces to the commutative triangle

$$\begin{array}{ccc} u_i(A) & \xrightarrow{\beta^{\mathcal{U}}_{u_j(A)}} & u_j(A) \\ & {\scriptstyle\beta_A^{\mathcal{U}_i}}\searrow \quad \swarrow{\scriptstyle\beta_A^{\mathcal{U}_j}} & \\ & A. & \end{array}$$

So we obtain a unique factorization

$$\beta_A^{\mathcal{U}} : u_! u^*(A) = u(A) \rightarrow A$$

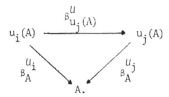

of the cone $(\beta_A^{u_i})_{i\in I}$ through the limit-cone $(s_i : u_i(A) \to u(A))_{i\in I}$. In order to prove that β^u is a natural transformation, we need some preliminary results.

Consider some index k in I. The functor $u_{k!}\, u_k^*$ has a right adjoint $u_{k*}\, u_k^*$, so it is cocontinuous. Applying this cocontinuous functor to the limit cone $(s_i : u_i(A) \to u(A))_{i\in I}$, we obtain a limit cone

$$(u_k(s_i) : u_k\, u_i(A) \to u_k\, u(A))_{i\in I}.$$

But we know that $u_k\, u_i(A)$ is just $(u_k \wedge u_i)\ (A)$. So the last limit cone is defined on a diagram with a terminal object $u_k(A)$ (take $i = k$) which therefore is its colimit. This gives the isomorphisms

$$u_k(uA) \cong u_k(A)$$

$$u_k(s_i) \cong \beta_{u_kA}^{u_i}.$$

In particular we have

$$\begin{aligned} s_k &= s_k \circ \beta_{u_kA}^{u_k} \\ &= \beta_{uA}^{u_k} \circ u_k(s_k) \\ &= \beta_{uA}^{u_k} \circ \beta_{u_kA}^{u_k} \\ &= \beta_{uA}^{u_k}. \end{aligned}$$

We willprove now that u is an idempotent endofunctor on C.

$$\begin{aligned} u\, u(A) &= \varinjlim_{i\in I} u_i(uA) \\ &= \varinjlim_{i\in I} u_i(A) \\ &= u(A). \end{aligned}$$

Moreover if $f : A \to B$ is a morphism in C, $u_k(uf)$ is the unique morphism which makes the following diagram commute :

$$\begin{array}{ccc} uA & \xrightarrow{\quad uf \quad} & uB \\ {\scriptstyle \beta^{\mathcal{U}_k}_{uA}} \uparrow & & \uparrow {\scriptstyle \beta^{\mathcal{U}_k}_{uB}} \\ u_k(uA) & \xrightarrow[\quad u_k(uf) \quad]{} & u_k(uB). \end{array}$$

The definition of uf and the relations we have just proved show that $u_k(f)$ makes the same diagram commutative : therefore $u_k(uf) = u_k(f)$. This implies that

$$\begin{aligned} u(uf) &= \varinjlim_{i \in I} u_i(uf) \\ &= \varinjlim_{i \in I} u_i(f) \\ &= uf. \end{aligned}$$

So u is an idempotent endofunctor on C. Moreover $\beta^{\mathcal{U}}_{uA} : u(uA) \to uA$ is the identity morphism because, if we compose it with the morphisms $\beta^{\mathcal{U}_i}_{uuA}$ of the limit cone, we have

$$\beta^{\mathcal{U}}_{uA} \circ \beta^{\mathcal{U}_i}_{uuA} = \beta^{\mathcal{U}_i}_{uA} = \beta^{\mathcal{U}_i}_{uuA}.$$

We are now able to prove that $\beta^{\mathcal{U}} : u \Rightarrow id_C$ is a natural transformation. If $f : A \to B$ is a morphism in C, we must prove that

$$f \circ \beta^{\mathcal{U}}_A = \beta^{\mathcal{U}}_B \circ u(f).$$

It suffices to compose each side with $\beta^{\mathcal{U}_i}_{uA}$:

$$\begin{aligned} \beta^{\mathcal{U}}_B \circ u(f) \circ \beta^{\mathcal{U}_i}_{uA} &= \beta^{\mathcal{U}}_B \circ \beta^{\mathcal{U}_i}_{uB} \circ u_i(f) \\ &= \beta^{\mathcal{U}_i}_B \circ u_i(f) \\ &= f \circ \beta^{\mathcal{U}_i}_A \\ &= f \circ \beta^{\mathcal{U}}_A \circ \beta^{\mathcal{U}_i}_{uA}. \end{aligned}$$

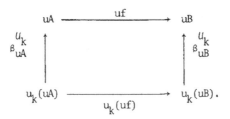

$u_!$ is a canonical inclusion and any morphism in $\mathcal{U}$ has the form $u^*(f)$ for some f in C; so the equality

$$u_!\ u^*\ u_!\ u^* = uu = u = u_!\ u^*$$

shows that in fact $u^*\ u_!$ is the identity on $\mathcal{U}$. So we have two natural transformations

$$id^{\mathcal{U}} : id_{\mathcal{U}} \Longrightarrow u^*\ u_!$$

$$\beta^{\mathcal{U}} : u_!\ u^* \Rightarrow id_C .$$

In order to have an adjunction $u_! \dashv u^*$, it remains to show that the compatibility conditions hold :

$$u^* * \beta^{\mathcal{U}} = id_{u^*}$$

$$\beta^{\mathcal{U}} * u_! = id_{u_!} .$$

Let A be some object in C. We first check that for any i in I, $u_i(\beta^{\mathcal{U}}_A)$ is the identity on $u_i(A)$. Now $u_i(\beta^{\mathcal{U}}_A)$ is the unique morphism which makes the following diagram commute

$$\begin{array}{ccc} uA & \xrightarrow{\ \beta^{\mathcal{U}}_A\ } & A \\ {\scriptstyle \beta^{\mathcal{U}_i}_{uA}}\uparrow & & \uparrow{\scriptstyle \beta^{\mathcal{U}_i}_{A}} \\ u_i(uA) & \xrightarrow[\ u_i(\beta^{\mathcal{U}}_A)\ ]{} & u_i(A) \end{array}$$

But we know that the identity on $u_i(A)$ makes this diagram commute; so $u_i(\beta^{\mathcal{U}}_A)$ is the identity morphism. On the other hand, $u(\beta^{\mathcal{U}}_A)$ is the unique morphism such that for any i

$$u(\beta^{\mathcal{U}}_A) \circ \beta^{\mathcal{U}_i}_{uuA} = \beta^{\mathcal{U}_i}_{uA} \circ u_i(\beta^{\mathcal{U}}_A)$$

which means exactly that

$$u(\beta^{\mathcal{U}}_A) \circ \beta^{\mathcal{U}_i}_{uA} = \beta^{\mathcal{U}_i}_{uA} .$$

The identity on uA is clearly such a morphism and $u(\beta^{\mathcal{U}}_A) = id_{uA}$. This proves the first condition for the adjunction. The second is exactly the equality $\beta^{\mathcal{U}}_{uA} = id_{uA}$ which has already been proved. So $u_!$ is left adjoint to u^*.

We have to verify that $u_!$ and u^* satisfy F 3 - F 4 - F 5. We first prove that the image of $\beta_A^{\mathcal{U}}$, for any A in C, is the union of the images of the various $\beta_A^{\mathcal{U}_i}$. Consider the following diagram where $\beta_A^{\mathcal{U}}$ and $\beta_A^{\mathcal{U}_i}$ have been factored through their image :

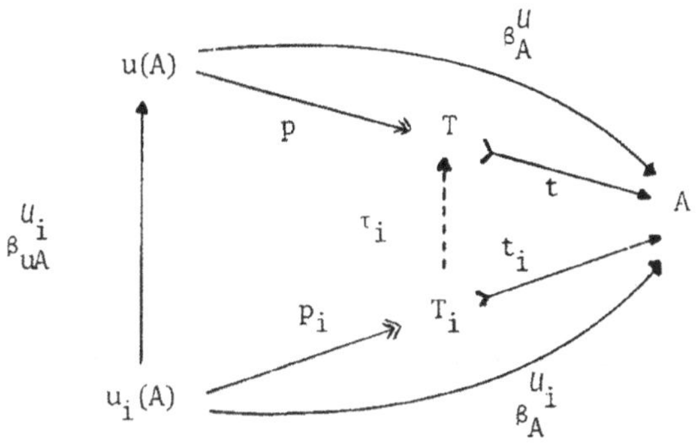

There is a factorization τ_i through the images. Thus we obtain the following commutative diagram for $\mathcal{U}_i \leqslant \mathcal{U}_j$

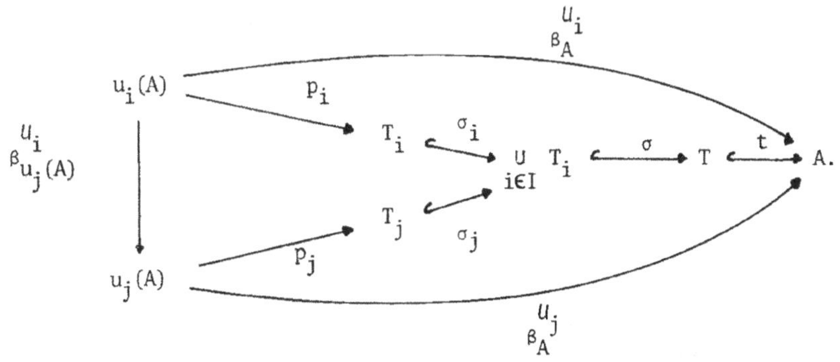

So we obtain a cone with vertex $\bigcup_{i\in I} T_i$ and thus a unique factorization

$$q : u(A) \to \bigcup_{i\in I} T_i \text{ such that } q \circ \beta_{uA}^{\mathcal{U}_i} = \sigma_i \circ p_i$$

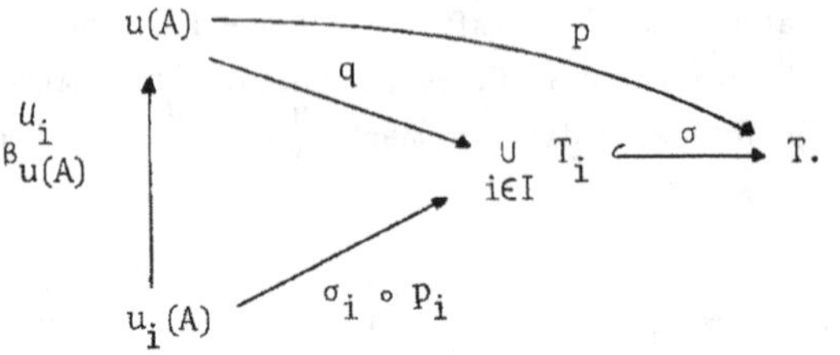

But $p \circ \beta^{\mathcal{U}_i}_{uA} = \tau_i \circ p_i$

$$= \sigma \circ \sigma_i \circ p_i$$

$$= \sigma \circ q \circ \beta^{\mathcal{U}_i}_{uA}$$

and thus $p = \sigma \circ q$. p is a regular epimorphism and so the same holds for the monomorphism σ. Hence σ is an isomorphism. Thus $T = \bigcup_{i \in I} T_i$.

Suppose $0 \to A$ is a monomorphism in C. Then $\beta^{\mathcal{U}}_A$ is the filtered union of the monomorphisms $\beta^{\mathcal{U}_i}_A$. Thus it is a monomorphism (proposition I - 10). Moreover it is a Heyting subobject since it is a union of Heyting subobjects (theorem 4). So F 4 and F 5 hold. Now if $0 \to M$ and $f : A \to M$ are monomorphisms in C with M an object in $\mathcal{U}$, we have $M = uM$ and thus

$$A = A \cap M$$

$$= A \cap u(M)$$

$$= A \cap \left(\bigcup_{i \in I} u_i(M) \right)$$

$$= \bigcup_{i \in I} (A \cap u_i(M)) \qquad \text{by I H 2}$$

$$= \bigcup_{i \in I} u_i(A) \qquad \text{proposition 8}$$

$$= u(A). \qquad 0 \to A \text{ is mono}$$

So A is in $\mathcal{U}$ and f is in $\mathcal{U}$.

We define now u_* right adjoint to u^*. If M is some object in $\mathcal{U}$, we define

$$u_* M = \varinjlim_{i \in I} u_{i_*} u_i^* u_! M$$

where the limit is taken on the diagram with vertices $u_{i_*} u_i^* M$ and with the arrows $\alpha^{U_i}_{u_{j_*} u_j^* M} = u_{j_*} u_j^* (\alpha_M^{U_i}) : u_{j_*} u_j^* M \to u_{i_*} u_i^* M$ given by corollary 10 for $U_i \leqslant U_j$. If $m : M \to N$ is a morphism in U, the naturality of the $\alpha_M^{U_i}$ implies that the morphisms

$$u_{i_*} u_i^* f : u_{i_*} u_i^* M \to u_{i_*} u_i^* N$$

determine a natural transformation between the diagrams defining $u_* M$ and $u_* N$; so there is a unique factorization $u_* m : u_* M \to u_* N$ and u_* becomes a functor from U to C.

The following natural bijections show that u_* is right adjoint to u^*. For any A in C and M in U one has

$$\begin{aligned}
(u^* A, M) &\cong (u_! u^* A, u_! M) \\
&\cong (\varinjlim_{i \in I} u_{i_!} u_i^* A, u_! M) \\
&\cong \varprojlim_{i \in I} (u_{i_!} u_i^* A, u_! M) \\
&\cong \varprojlim_{i \in I} (u_i^* A, u_i^* u_! M) \\
&\cong \varprojlim_{i \in I} (A, u_{i_*} u_i^* u_! M) \\
&\cong (A, \varprojlim_{i \in I} u_{i_*} u_i^* u_! M) \\
&\cong (A, u_* M).
\end{aligned}$$

But we have to prove that the last limit is computed on the diagram defining $u_* M$ as a colimit. By the definition of uA, the limit

$$\varprojlim_{i \in I} (u_{i!} u_i^* A, M)$$

is computed on the diagram with arrows

$$(u_{j!} u_j^* A, M) \xrightarrow{(\beta^{U_i}_{u_{j!} u_j^* A}, 1)} (u_{i!} u_i^* A, M)$$

for $U_i \leqslant U_j$. Applying the bijections we obtain an arrow

$$(A, u_{j_*} u_j^* M) \longrightarrow (A, u_{i_*} u_i^* M)$$

wich may be described as follows : a map

$$f : A \to u_{j_*} u_j^* M$$

corresponds in $(u_{j!} u_j^* A, M)$ to the composite

$$u_{j!} u_j^* A \xrightarrow{u_{j!} u_j^* f} u_{j!} u_j^* u_{j_*} u_j^* M = u_{j!} u_j^* M \xrightarrow{\beta_M^{U_j}} M$$

and is carried in $(u_{i!} u_{i_*} A, M)$ to the composite

$$\beta_M^{U_j} \circ u_{j!} u_j^* f \circ \beta^{U_i}_{u_{j!} u_j^* A}.$$

By naturality this composite corresponds in $(A, u_{i_*} u_i^* M)$ to

$$u_{i_*} u_i^* \beta_M^{U_j} \circ u_{i_*} u_i^* u_{j!} u_j^* f \circ u_{i_*} u_i^* \beta^{U_i}_{u_{j!} u_j^* A} \circ \alpha_A^{U_i}$$

$$= u_{i_*} u_i^* f \circ u_{i_*} u_i^* \beta_A^{U_j} \circ u_{i_*} u_i^* \beta^{U_i}_{u_{j!} u_j^* A} \circ \alpha_A^{U_i}$$

$$= u_{i_*} u_i^* f \circ \alpha_A^{U_i} = \alpha^{U_i}_{u_{j_*} u_j^* A} \circ f.$$

($u_i^* \beta_A^{U_j}$ and $u_i^* \beta^{U_i}_{u_{j!} u_j^* A}$ are identities; see corollary 10).

So the arrows in the diagram defining the last limit are

$$(A, u_{j_*} u_j^* M) \xrightarrow{(1, \alpha^{U_i}_{u_{j_*} u_j^* A})} (A, u_{i_*} u_i^* M).$$

This concludes the proof that u_* is right adjoint to u^* and finally that U is a formal initial segment in C.

We now claim that U is the supremum of the family $(U_i)_{i \in I}$. Let M be some object in some U_k. Then for any U_i such that $U_k \leq U_i$, M is in U_i and thus $u_i(M) \cong M$. Thus the diagram defining uM has a terminal object M and uM = M. So M is in U and $U_k \leq U$. Now consider V such that for any $i \in I$, $U_i \leq V$. Take M in U. Each $u_{i!} u_i^* M$ is in U_i and thus in V. So the colimit

$\varinjlim_{i\in I} u_i(M) = uM$ is in V ($v_!$ is cocontinuous and V is cocomplete). From this we deduce the inequality $U \leqslant V$. Thus U is the supremum of the family $(U_i)_{i\in I}$.

Finally, to verify that H is a frame, we must prove the infinite distributivity law

$$V \wedge (\bigvee_{i\in I} U_i) = \bigvee_{i\in I} (V \wedge U_i)$$

for initial segments V and U_i in C. But for U any formal initial segment, the subcategory U is exactly the image of the composite functor

$$C \xrightarrow{u^*} U \xrightarrow{u_!} C.$$

So it suffices to prove that

$$(v \wedge (\bigvee_{i\in I} u_i))_! \, (v \wedge (\bigvee_{i\in I} u_i))^*$$

$$= (\bigvee_{i\in I} (v \wedge u_i))_! \, (\bigvee_{i\in I} (v \wedge u_i))^*.$$

But all the functors involved have a right adjoint and thus are cocontinuous. So it suffices to prove the equality on a dense subcategory. The category of those objects A such that $0 \to A$ is a monomorphism is such a category (proposition I - 7). Now if $0 \to A$ is a monomorphism then

$$(u \wedge v)\ (A) = u(A) \cap v(A)$$

$$(\bigvee_{i\in I} u_i)\ (A) = \bigvee_{i\in I} u_i(A).$$

So it suffices to prove that

$$v\,A \cap (\bigcup_{i\in I} u_i\,A) = \bigcup_{i\in I} (v\,A \cap u_i\,A).$$

And this is true since all these subobjects are Heyting subobjects. ∎

§ 4. COMPARISON OF VARIOUS FRAMES RELATED TO Sh($\mathbb{H}$, $\mathbb{T}$)

Starting with a frame $\mathbb{H}$ and an algebraic theory $\mathbb{T}$ in Sh($\mathbb{H}$) we were able to construct the new frame H of formal initial segments of Sh($\mathbb{H}$, $\mathbb{T}$) without referring to the elements of $\mathbb{H}$. We will now establish that $\mathbb{H}$ is a subframe of H and that H itself is a subframe of the frame of Heyting subobjects of the free $\mathbb{T}$-algebra a F h_1.

Proposition 12.

Let H *be the frame of formal initial segments of* Sh($\mathbb{H}, \mathbb{T}$). *Let* *Heyt* (a F h_1) *be the frame of Heyting subobjects of* a F h_1. *There are inclusions of frames*

$$\mathbb{H} \subseteq H \subseteq \text{Heyt (a F } h_1).$$

In particular, H *is a set.*

By proposition I - 5 we know that the morphism $0 \to$ a F h_1 is a monomorphism.

We will first construct the inclusion :

$$H \hookrightarrow \text{Heyt (a F } h_1).$$

A formal initial segment U is sended to $u_! \, u^*$(a F h_1) which is a Heyting subobject of a F h_1 (by F 4 - F 5). If $U \leqslant V$ are formal initial segments, the relation u(a F h_1) $\leqslant$ v(a F h_1) holds (cfr. propositions 9 - 10). On the other hand, o(a F h_1) = 0 and i(a F h_1) = a F h_1 (cfr. theorem 11). Thus we have a morphism of frames

$$H \longrightarrow \text{Heyt (a F } h_1).$$

Now suppose we are given U and V formal initial segments in Sh($\mathbb{H}, \mathbb{T}$) such that u(a F h_1) is isomorphic to v(a F h_1) as subobjects of a F h_1, i.e. $\beta^U_{\text{a F } h_1} = \beta^V_{\text{a F } h_1}$. For any element w in $\mathbb{H}$ we have (proposition 8)

$$\begin{aligned} u(\text{a F } h_w) &= \text{a F } h_w \cap u(\text{a F } h_1) \\ &= \text{a F } h_w \cap v(\text{a F } h_1) \\ &= v(\text{a F } h_w) \end{aligned}$$

and $\beta^U_{\text{a F } h_w} = \beta^V_{\text{a F } h_w}$.

But $u_! \, u^*$ and $v_! \, v^*$ have a right adjoint and thus are cocontinuous; so they coincide on each coproduct $\coprod_{i\in I}$ a F h_{w_i} where $w_i \in \mathbb{H}$ and $\beta^U_{\coprod_{i\in I} \text{a F } h_{w_i}} = \beta^V_{\coprod_{i\in I} \text{a F } h_{w_i}}$.

Now if $g : \coprod_{i\in I} \text{a F } h_{w_i} \to \coprod_{j\in J} \text{a F } h_{w_j}$ is some morphism, the following relations hold

$$\beta^U_{\coprod_{j\in J} \text{a F } h_{w_j}} \circ u(f) = f \circ \beta^U_{\coprod_{i\in I} \text{a F } h_{w_i}}$$

$$= f \circ \beta \coprod_{i\in I}^{V} a\, F\, h_{w_i}$$

$$= \beta \coprod_{j\in J}^{V} a\, F\, h_{w_j} \circ v(f)$$

$$= \beta \coprod_{j\in J}^{U} a\, F\, h_{w_j} \circ v(f).$$

But $\beta \coprod_{j\in J}^{U} a\, F\, h_{w_j}$ is a monomorphism; so $u(f) = v(f)$.

Finally u and v coincide on a dense subcategory of Sh($\mathbb{H}$, $\mathbb{T}$) (cfr. proposition I - 7) and are cocontinuous and hence they are isomorphic (cfr. [21] - 17 - 2 - 7). We have already noted (proof of theorem 11) that this is sufficient to imply the isomorphism of U and V. So we have shown the injectivity of the morphism $H \to$ Heyt $(a\, F\, h_1)$.

On the other hand theorem 5 produces a mapping $\alpha : \mathbb{H} \to H$. The composition of this map α with the inclusion of frames $H \subseteq$ Heyt $(a\, F\, h_1)$ is simply the mapping wich sends u to $u(a\, F\, h_1) = a\, F\, h_u$. But for w in $\mathbb{H}$

$$F\, h_u(w) = \begin{cases} F\,1 & \text{if} \quad w \leqslant u \\ F\,0 & \text{if} \quad w \nleqslant u. \end{cases}$$

So for u and v in $\mathbb{H}$

$$F\, h_u \cap F\, h_v = F\, h_{u \wedge v}$$

and applying the associated sheaf functor

$$a\, F\, h_u \cap a\, F\, h_v = a\, F\, h_{u \wedge v}.$$

Now if $(u_i)_{i\in I}$ is a family of elements in $\mathbb{H}$ and $w \in \mathbb{H}$,

$$\bigcup F\, h_{u_i}(w) = \begin{cases} F\,1 & \text{if} \quad \exists i \quad w \leqslant u_i \\ F\,0 & \text{if} \quad \text{not}. \end{cases}$$

$$F\, h_{\bigcup_{i\in I} u_i}(w) = \begin{cases} F\,1 & \text{if} \quad w \leqslant \bigcup_{i\in I} u_i \\ F\,0 & \text{if} \quad \text{not}. \end{cases}$$

So $\bigcup F\, h_{u_i}$ is a sub-presheaf of $F\, h_{\bigcup_{i\in I} u_i}$. But if x is some element in

$\cup F h_{u_i}(w)$ with $w \leq \bigcup_{i\in I} u_i$, then $w = \bigcup_{i\in I} (w \wedge u_i)$ and the family $(x_{|w \wedge u_i})_{i\in I}$ is in the presheaf $F h_{\bigcup_{i\in I} u_i}$. So both presheaves have the same associated sheaf and $a F h_{\bigcup_{i\in I} u_i} = a(\bigcup_{i\in I} F h_{u_i}) = \bigcup_{i\in I} a F h_{u_i}$. It is also clear that $a F h_0 = 0$. So the mapping

$$\mathbb{H} \longrightarrow \text{Heyt } (a\ F\ h_1)$$

which sends u to $a F h_u$ is a morphism of frames.

We will now prove that this last mapping is injective. Consider u and v in $\mathbb{H}$ such that $a F h_u$ is isomorphic to $a F h_v$. It then follows that

$$a\ F\ h_{u \wedge v} \cong a\ F\ h_u \wedge a\ F\ h_v \cong a\ F\ h_u$$

and this allows us to suppose that $u \leq v$. If x is the universal generator of $F h_v(v)$, x is some element in $a F h_v(v) = a F h_u(v)$. So there exists a family $(v_i)_{i\in I}$ with $v = \bigvee_{i\in I} v_i$ such that $x_{|v_i} \in F h_u(v_i)$. But $x_{|v_i}$ is the universal generator in $F h_v(v_i)$ and thus $F h_u(v_i) = F h_v(v_i)$; in other words, $v_i \leq u$. This is true for any i and thus $v = \bigvee_{i\in I} v_i \leq u$, which proves the equality $u = v$.

Finally $\mathbb{H}$ is a subframe of Heyt $(a\ F\ h_1)$ and the canonical inclusion factors through H. This concludes the proof. ■

§ 5. SHEAVES ON THE FRAME OF FORMAL INITIAL SEGMENTS

In this paragraph we limit our investigation to the case of a classical theory $\mathbb{T}$. In this case both $Sh(\mathbb{H}, \mathbb{T})$ and $Sh(H, \mathbb{T})$ make sense and are the sheaves on $\mathbb{H}$ and H with values in $\text{Sets}^{\mathbb{T}}$. From the inclusion $\mathbb{H} \hookrightarrow H$ (proposition 12) we deduce by composition a restriction functor

$$\Gamma : Sh(H, \mathbb{T}) \to Sh(\mathbb{H}, \mathbb{T}).$$

We will prove that this restriction functor has a left exact left adjoint Θ (a kind of "algebraic geometric morphism"). But Γ also has a right inverse Δ. The particularization of Δ to the case of modules on a ring will produce the theorems of further chapters on sheaf-representation of rings and modules.

Proposition 13.

Let $\mathbb{H}$ be a frame, $\mathbb{T}$ a classical theory and H the frame of formal initial segments in $Sh(\mathbb{H}, \mathbb{T})$. The restriction functor $\Gamma : Sh(H, \mathbb{T}) \to Sh(\mathbb{H}, \mathbb{T})$

has an exact left adjoint Θ.

We first define a functor

$$\Theta' : Sh(\mathbb{H}, \mathbb{T}) \to Sh(H, \mathbb{T}).$$

If A is some object in $Sh(\mathbb{H}, \mathbb{T})$ and $\mathcal{U}$ some formal initial segment, define

$$\Theta' A(\mathcal{U}) = \varinjlim_{\substack{v \geq \mathcal{U} \\ v \in \mathbb{H}}} A(v).$$

Clearly if $\mathcal{U} \leq \mathcal{U}'$ are formal initial segments and $v \in \mathbb{H}$, $v \geq \mathcal{U}'$; then $v \geq \mathcal{U}$. This produces a canonical factorization $\Theta' A(\mathcal{U}') \to \Theta' A(\mathcal{U})$ which makes $\Theta' A$ into a presheaf on H. In the same way if $f : A \to B$ is a morphism in $Sh(\mathbb{H}, \mathbb{T})$, the morphisms $f_v : A(v) \to B(v)$ induce a canonical factorization $\Theta' A(\mathcal{U}) \to$ $\to \Theta' B(\mathcal{U})$ and finally a morphism $\Theta'(A) \to \Theta'(B)$; this makes Θ' into a functor. Θ is the composite of Θ' and the associated sheaf functor.

To prove the adjunction, consider a sheaf F in $Sh(H, \mathbb{T})$. We need to produce a morphism γ_F

$$\gamma_F : \Theta \Gamma F = a\, \Theta' \Gamma F \to F$$

which is equivalent, by adjunction, to a morphism

$$\gamma'_F : \Theta' \Gamma F \to F$$

So, if $\mathcal{U}$ is some formal initial segment, we must find a homomorphism

$$\gamma'_F(\mathcal{U}) : \varinjlim_{\substack{v \geq \mathcal{U} \\ v \in \mathbb{H}}} F(v) \to F\ \mathcal{U};$$

Take it to be the factorization through the colimit of the restriction map $F(v) \to F\ \mathcal{U}$. Clearly γ'_F is a natural transformation, i.e. a morphism of presheaves.

Now consider A in $Sh(\mathbb{H}, \mathbb{T})$ and a morphism $f : \Theta A \to F$ in $Sh(H, \mathbb{T})$. We must find a unique factorization $g : A \to \Gamma F$ such that $\gamma_F \circ \Theta(g) = f$.

$$\begin{array}{ccc} \Theta \Gamma F = a\, \Theta' \Gamma F & \xrightarrow{\ \gamma_F\ } & F \\ \Theta g \Big\uparrow a\, \Theta' g & \nearrow_{f} & \\ \Theta A = a\, \Theta' A & & \end{array}$$

If such a g exists, evaluate this diagram at $v \in \mathbb{H}$ and get

$$\begin{array}{ccc} Fv & = & Fv \\ {\scriptstyle g}\uparrow & \nearrow {\scriptstyle f_v} & \\ A(v) & & \end{array} \quad .$$

This shows uniqueness of g. To prove the existence, consider then the morphism $g : A \to \Gamma F$ defined by $g_v = f_v$ for any v in $\mathbf{H}$. If U is some formal initial segment, we need to prove the commutativity of

$$\begin{array}{ccc} \varinjlim_{\substack{v \geq U \\ v \in \mathbf{H}}} F(v) & \xrightarrow{\gamma_F(U)} & F(U) \\ {\scriptstyle \varinjlim_{\substack{v \geq U \\ v \in H}} f_v}\uparrow & \nearrow {\scriptstyle f(U)} & \\ \varinjlim_{\substack{v \geq U \\ v \in \mathbf{H}}} A(v) & & \end{array}$$

If we compose the diagram with any canonical morphism

$$s_v^A : A(v) \to \varinjlim_{\substack{v \geq U \\ v \in \mathbf{H}}} A(v)$$

we obtain the following equalities :

$$\begin{aligned} \gamma_F(U) \circ \varinjlim_{\substack{v \geq U \\ v \in \mathbf{H}}} f_v \circ s_v^A &= \gamma_F(U) \circ s_v^F \circ f_v \\ &= F(U \leq v) \circ f_v \\ &= f_U \circ \Theta\, A(U \leq v) \\ &= f_U \circ s_v^A . \end{aligned}$$

By the universal property of colimits, we obtain the required equality. So Θ is left adjoint to Γ.

Now Θ' is defined by a filtered colimit. But in $Sh(\mathbf{H}, \mathbb{T})$ and $Sh(H, \mathbb{T})$ filtered colimits commute with finite limits (proposition I - 3). So Θ' commutes with finite limits. But the associated sheaf functor also commutes with finite limits. Thus Θ is exact. ∎

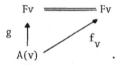

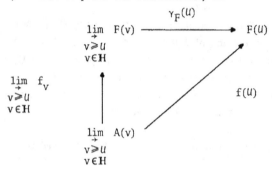

Proposition 14.

Let $\mathbb{H}$ *be a frame,* $\mathbb{T}$ *a classical theory and* $\mathcal{H}$ *the frame of formal initial segments in* $Sh(\mathbb{H},\mathbb{T})$. *The restriction functor* $\Gamma : Sh(\mathcal{H},\mathbb{T}) \to Sh(\mathbb{H},\mathbb{T})$ *has a right inverse* Δ *which is continuous and faithful.*

If A is some object in $Sh(\mathbb{H},\mathbb{T})$ and $\mathcal{U}$ a formal initial segment, we define Δ A by

$$\Delta A(\mathcal{U}) = (u_* u^* A)(1).$$

If $\mathcal{U} \leq \mathcal{V}$ are two formal initial segments, the restriction morphism

$$\Delta A(\mathcal{U} \leq \mathcal{V}) : (v_* v^* A)(1) \to (u_* u^* A)(1)$$

is the one given by proposition 10.

$$\Delta A(\mathcal{U} \leq \mathcal{V}) = v_* v^*(\alpha_A^{\mathcal{U}})(1) = \alpha^{\mathcal{U}}_{v_* v^* A}(1).$$ This makes Δ A into a presheaf.

Now if $f : A \to B$ is some morphism in $Sh(\mathbb{H},\mathbb{T})$, the morphism $\Delta(f)$ is defined for any formal initial segment $\mathcal{U}$ by

$$\Delta f(\mathcal{H}) = (u_* u^* f)(1).$$

This is clearly a morphism of presheaves and Δ becomes a functor

$$\Delta : Sh(\mathbb{H},\mathbb{T}) \to Pr(\mathcal{H},\mathbb{T}).$$

A remarkable fact, which will be crucial for the sheaf-representation theorems, is that each Δ A is actually a sheaf. Indeed, let $\mathcal{U} = \bigvee_{i\in I} \mathcal{U}_i$ in $\mathcal{H}$. We know we may suppose the family $(\mathcal{U}_i)_{i\in I}$ to be stable by finite intersections. We have

$$\Delta A(\mathcal{U}) = (u_* u^* A)(1) = \varprojlim_{i\in I} (u_{i_*} u_i^* A)(1).$$

But an element in this (filtered) colimit is exactly a compatible family $(x_i)_{i\in I}$ of elements choosen in $(u_{i_*} u_i^* A)(1) = \Delta(A)(\mathcal{U}_i)$. Thus Δ A is a sheaf on $\mathcal{H}$.

We prove now that Δ is right inverse to Γ. For any object A in $Sh(\mathbb{H},\mathbb{T})$ and u in $\mathbb{H}$

$$
\begin{aligned}
(\Gamma \, \Delta \, A)(u) &\cong \Delta \, A(u) \\
&\cong (u_* u^* A)(1) \\
&\cong (a \, F \, h_1, u_* u^* A) && \text{prop. I - 5} \\
&\cong (u^* a \, F \, h_1, u^* A) && u^* \dashv u_*
\end{aligned}
$$

$$\cong (u_! \; u^* \; a \; F \; h_1, A) \qquad u_! \dashv u^*$$

$$\cong (a \; F \; h_u, A) \qquad \text{theorem 5}$$

$$\cong A(u) \qquad \text{prop. I - 5}$$

Now if $u \leqslant v$, we will establish that

$$(\Gamma \; \Delta \; A)(u \leqslant v) \cong A(u \leqslant v)$$

using the isomorphisms described above. Indeed, the naturality of these isomorphisms implies

$$
\begin{aligned}
(\Gamma \; \Delta \; A)(u \leqslant v) &= (\Delta \; A)(u \leqslant v) \\
&= v_* \; v^* (\alpha_A^u)(1) \\
&\cong (1, \; v_* \; v^* \; \alpha_A^u) \\
&\cong (1, \; v^* \; \alpha_A^u) \\
&\cong (1, \; \alpha_A^u) \\
&\cong (1, \; \alpha_A^u) \\
&\cong \alpha_A^u(v).
\end{aligned}
$$

But $\alpha_A^u(u) : A(v) \to (u_* \; u^* \; A)(v) = A(u)$ is exactly the restriction morphism $A(u \leqslant v)$, by definition of u_*. This concludes the proof that $\Gamma \; \Delta \; A$ is isomorphic to A.

Finally, if $f : A \to B$ is some morphism in $Sh(\mathbb{H}, \mathbb{T})$, the naturality of the same isomorphisms implies, for u in $\mathbb{H}$

$$
\begin{aligned}
(\Gamma \; \Delta \; f)(u) &= (\Delta \; f)(u) \\
&= (u_* \; u^* \; f)(1) \\
&\cong (1, \; u_* \; u^* \; f) \\
&\cong (1, \; u^* \; f) \\
&\cong (1, \; f) \\
&\cong f(u).
\end{aligned}
$$

Thus we have proved the isomorphism $\Gamma \; \Delta \cong \text{id}$.

From the isomorphism $\Gamma \; \Delta \cong \text{id}$ and the fact that id is faithful, we deduce that Δ is faithful. Now limits are computed pointwise in $Sh(\mathbb{H}, \mathbb{T})$ and

$Sh(H, \mathbb{T})$ and limits are preserved by u_* and u^* which have a left adjoint; this proves that Δ is continuous. ■

CHAPTER 3 : LOCALIZATIONS AND ALGEBRAIC SHEAVES

Throughout this chapter, $\mathbb{H}$ is a fixed frame and $\mathbb{T}$ is a fixed finitary algebraic theory internally defined with respect to the topos $Sh(\mathbb{H},\mathbb{T})$. We intend to classify the localizations of $Sh(\mathbb{H},\mathbb{T})$ by means of algebraic Lawvere - Tierney topologies and by means of algebraic Gabriel - Grothendieck topologies.

A localization of $Sh(\mathbb{H},\mathbb{T})$ is a full reflexive subcategory of $Sh(\mathbb{H},\mathbb{T})$ whose reflexion is exact. For example, if $\mathbb{T}$ is the initial theory, $Sh(\mathbb{H},\mathbb{T})$ is just the topos $Sh(\mathbb{H})$ and a localization of $Sh(\mathbb{H})$ is exactly a subtopos of $Sh(\mathbb{H})$, i.e. a topos of sheaves for some Lawvere - Tierney topology on the Ω-object of $Sh(\mathbb{H})$ (cfr. [12]). Another example : if $\mathbb{H}$ is the initial frame $\{0, 1\}$ and $\mathbb{T}$ is the theory of modules on a ring R, a localization of $Sh(\mathbb{H},\mathbb{T}) = \underline{Mod}_R$ is exactly a "Gabriel localization of $\underline{Mod}_R$", i.e. the localization of $\underline{Mod}_R$ for some localizing system of ideals on R (cfr. [6] or [20]).

The definition of a "Gabriel - Grothendieck"-topology on $Sh(\mathbb{H},\mathbb{T})$ can be obtained in a straightforward way from the classical notion. An algebraic crible is any subobject J of some $a\,F\,h_u$ in $Sh(\mathbb{H},\mathbb{T})$. An algebraic Gabriel - Grothendieck topology is defined by specifying for each u in $\mathbb{H}$, a family of subobjects of $a\,F\,h_u$ which satisfy the three classical axioms and a fourth axiom which takes into account the topology already existing on $\mathbb{H}$. This fourth axiom does not appear in the classical case because a Gabriel - Grothendieck topology is defined on a small category and intends to classify the localizations of a category of presheaves. Here we define a topology on a site $\mathbb{H}$ and we intend to classify the localizations of some category of sheaves.

The definition of a "Lawvere - Tierney"-topology on $Sh(\mathbb{H},\mathbb{T})$ is less straightforward. From the category $Sh(\mathbb{H},\mathbb{T})$ we construct a topos $\mathcal{E}(\mathbb{H},\mathbb{T})$ and an object $\Omega_{\mathbb{T}}$ in this topos. Generally, $\Omega_{\mathbb{T}}$ is not the classifying object of $\mathcal{E}(\mathbb{H},\mathbb{T})$ but it has the structure of an inf-semi-lattice with greatest element. So it makes sense to consider a "Lawvere - Tierney"-topology on $\Omega_{\mathbb{T}}$, i.e. a morphism $\Omega_{\mathbb{T}} \to \Omega_{\mathbb{T}}$ which satisfies the three classical axioms. This is what we call an algebraic Lawvere - Tierney topology.

The main result of this chapter shows that, when the theory $\mathbb{T}$ is commutative, there is a one-to-one correspondance between the localizations of $Sh(\mathbb{H},\mathbb{T})$, the algebraic Gabriel - Grothendieck topologies on Sh(H, T) and the algebraic Lawvere - Tierney topologies on Sh(H, T). To prove this theorem, it is useful to point out that $\Omega_{\mathbb{T}}$ classifies in some sense the subobjects in Sh(H, T).

For the sake of clarity, we suppose immediately and through the whole chapter that T is a *commutative* theory internally defined with respect to the topos Sh(H) of sheaves on the frame H.

§ 1. SOME TECHNICAL LEMMAS

Let H be a frame and T a commutative theory internally defined in the topos Sh(H). Sh(H, T) is therefore a symmetric monoïdal closed category.

Lemma 1.

For any u *in* H *and* A *in* Sh(H, T), *the morphisms* $0 \to A$ *and* $\beta_A : uA \to A$ *are monomorphisms, where* β_A *is the universal morphism arising from the adjunction* $u_! \dashv u^*$.

The commutativity condition on a classical theory implies that two arbitrary constants must be equal. Thus a commutative theory has none or a single constant. But any mapping with domain the empty set or the singleton is injective. This proves that for any B in Pr(H, T), the morphism $0 \to B$ is a monomorphism. In particular for A in Sh(H, T) the morphism $0 \to A$ is a monomorphism in Pr(H, T); applying the associated sheaf functor, we obtain a monomorphism $0 \to A$ in Sh(H, T). We conclude by theorem I - 5. ∎

For any u in H and A in Sh(H, T), u^*A is the restriction of A to $u{\downarrow}$. But by way of $u_!$, we may look at Sh($u{\downarrow}$, T) as a subcategory of Sh(H, T) so that $u_!\, u^* A$ can be thought of as a subobject of A obtained by restricting A to $u{\downarrow}$. Lemma 1 makes this interpretation valid. So, at this point we shall find it convenient to introduce the notation $A|_u$ to denote the subobject $\beta_A^u : u_!\, u^*(A) \to A$. Similarly, for f a morphism in Sh(H, T), $u_!\, u^*(f)$ will be denoted by $f|_u$.

Lemma 2.

Let $f : A \to B$ *be a morphism in* Sh(H, T) *and* $1 = \bigvee_{i \in I} u_i$ *in* H.

Then $$A = \bigcup_{i \in I} f^{-1}(B|_{u_i}).$$

In particular, $$B = \bigcup_{i \in I} B|_{u_i}.$$

Consider the following diagram where the square is a pullback and g is the unique factorization :

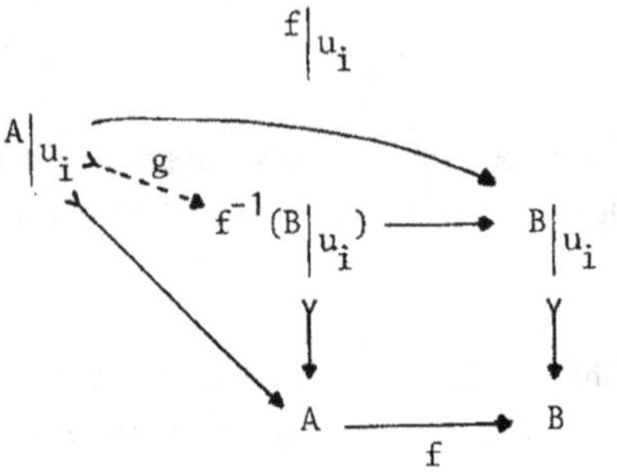

$$
\begin{aligned}
A &= A|_1 \\
&= A|_{\bigvee_{i\in I} u_i} \\
&= \bigcup_{i \in I} A|_{u_i} \qquad \text{(theorem II - 11).} \\
&\subseteq \bigcup_{i \in I} f^{-1}(B|_{u_i}).
\end{aligned}
$$

This proves the first assertion. The second assertion follows from the first one by taking $f = id_B$. ■

Lemma 3.

If $A' \rightarrowtail A$ *is some subobject in* $Sh(\mathbb{H}, \mathbb{T})$ *and* u *some element in* $\mathbb{H}$, $A'|_u = A' \cap A|_u$.

This is just proposition II - 8. ■

Lemma 4.

Let $f_1, \ldots, f_n : A \to B$ *be morphisms in* $Sh(\mathbb{H}, \mathbb{T})$ *and* $\alpha \in O_n(1)$ *some n-ary operation of* $\mathbb{T}(1)$. *Denote by* $f : A \to B$ *the morphism* $\alpha(f_1, \ldots, f_n)$. *Then for any subobject* $B' \rightarrowtail B$ *in* $Sh(\mathbb{H}, \mathbb{T})$ *we have the inclusion*

$$\bigcap_{i=1}^{n} f_i^{-1}(B') \subseteq f^{-1}(B').$$

The commutativity of the theory implies that $\alpha(f_1, \ldots, f_n)$ is a morphism in $Sh(\mathbb{H}, \mathbb{T})$ which is a closed category. But limits in $Sh(\mathbb{H}, \mathbb{T})$ are computed pointwise, so it suffices to prove the result in any commutatively algebraic category.

Consider x some element in $\bigcap_{i=1}^{n} f_i^{-1}(B')$. For any i, $f_i(x) \in B'$ and thus

$$f(x) = \alpha(f_1(x), \ldots, f_n(x)) \in B'$$

which shows that x is in $f^{-1}(B')$. ■

Lemma 5.

For any u *in* $\mathbb{H}$, *the functor* $u_!$ *commutes with non empty finite limits. In particular the restriction functor* $Sh(\mathbb{H}, \mathbb{T}) \to Sh(\mathbb{H}, \mathbb{T})$; $A \mapsto A|_u$ *commutes with non empty finite limits.*
If each $O_0(u)$ *is a singleton, this lemma extends to all finite limits.*

Consider two objects M, N in $\mathcal{U}$ and their product $M \times N$ in $\mathcal{U}$.

$$u'(M \times N)(v) = \begin{cases} M(v) \times N(v) & \text{if} \quad v \leqslant u \\ F0 & \text{if} \quad v \nleqslant u \end{cases}$$

$$(u'M \times u'N)(v) = \begin{cases} M(v) \times N(v) & \text{if} \quad v \leqslant u \\ F0 \times F0 & \text{if} \quad v \nleqslant u. \end{cases}$$

But the commutativity of $\mathbb{T}$ implies that F0 is the empty set or a singleton. Thus $F0 \times F0 = F0$. So $u'(M \times N) \cong u'M \times u'N$. Applying the associated sheaf functor, one gets $u_!(M \times N) \cong u_!M \times u_!N$. Now u^* has a left adjoint $u_!$ and is thus continuous; therefore u commutes with binary products.

Consider two morphisms $m, n : M \rightrightarrows N$ in $\mathcal{U}$ and their equalizer $k : K \to M$. For any $v \leqslant u$ in $\mathbb{H}$,

$$u'(K)(v) = K(v) = \mathrm{Ker}(u'm, u'n)(v).$$

Now if $v \not\leqslant u$ in $\mathbb{H}$,

$$u'M(v) = u'N(v) = FO = \{*\} \text{ or } u'M(v) = u'N(v) = FO = \phi.$$

But in both cases, we necessarily have

$$\mathrm{Ker}(u'm, u'n)(v) = FO = u'K(v).$$

This shows that $u_!$ and thus the restriction functor commute with equalizers.

Now let $\mathbb{T}$ be a theory such that each $\mathcal{O}_0(u)$ is a singleton; $u_!$ and the restriction functor $u_! u^*$ commute with finite limits. Indeed, it remains to show that $u_!$ commutes with the terminal object. But $u'(1)$ is the constant presheaf on 1, i.e. the object 1 in $\mathrm{Pr}(\mathbb{H}, \mathbb{T})$; therefore $u_!(1) = 1$ by exactness of the associated sheaf functor. ∎

Lemma 6.

If A *is some object in* Sh($\mathbb{H}, \mathbb{T}$) *and* α *some operation in* $\mathcal{O}_n(1)$, *there is a morphism* $\alpha_A : A^n \to A$ *in* Sh($\mathbb{H}, \mathbb{T}$) *such that for any morphisms* $f_1, \ldots, f_n : B \to A$ *in* Sh($\mathbb{H}, \mathbb{T}$), *one has*

$$\alpha(f_1, \ldots, f_n) = \alpha_A \circ (f_1, \ldots, f_n).$$

This follows easily from the fact that Sh($\mathbb{H}, \mathbb{T}$) is a closed category (cfr. [15]). α_A is just $\alpha(p_1, \ldots, p_n)$ where $p_i : A^n \to A$ is the i-th projection. ∎

In general we will simply use the notation α instead of α_A. The reader will point out that for an operation $\alpha \in \mathcal{O}_n(u)$ and for $u \in \mathbb{H}$, we can apply lemma 6 to Sh($u{\downarrow}$, $\mathbb{T}$) and A in Sh($\mathbb{H}, \mathbb{T}$) in order to obtain a morphism

$$\alpha : (u^*A)^n \to u^*A$$

in Sh($u{\downarrow}$, $\mathbb{T}$). If n is not zero, we obtain by lemma 5 a morphism

$$\alpha : A^n|_u \longrightarrow A|_u.$$

We shall refer to this remark as lemma 6 applied to Sh($u{\downarrow}$, $\mathbb{T}$).

Lemma 7.

For any v *in* $\mathbb{H}$ *there are isomorphisms*

$$\mathcal{O}_1(v) \cong (a\,F\,h_v,\ a\,F\,h_v).$$

Under these isomorphisms, the restriction of $x \in \mathcal{O}_1(v)$ *to* $u \leqslant v$ *corresponds* *to the application of the restriction functor* $u_! \, u^*$.

We know that $\mathcal{O}_1 = a\,F\,h_1$ and $h_1(w) = h_v(w) = 1$ for any $w \leqslant v$. Therefore we have (proposition I - 5)

$$\mathcal{O}_1(v) = a\,F\,h_1(v) \cong a\,F\,h_v(v) = (a\,F\,h_v,\ a\,F\,h_v).$$

In the same way $h_v(w) = h_u(w)$ for any $w \leqslant u$; therefore $a\,F\,h_v(u) = a\,F\,h_u(u)$. So, applying the restriction map

$$\mathcal{O}_1(u \leqslant v) = a\,F\,h_1(u \leqslant v) = a\,F\,h_v(u \leqslant v)$$

corresponds to composition with

$$a\,F\,h(u \leqslant v) = \beta^u_{a\,F\,h_v} : a\,F\,h_u \to a\,F\,h_v.$$

But for any g in $(a\,F\,h_v,\ a\,F\,h_v)$ the commutativity of the diagram

$$\begin{array}{ccc}
a\,F\,h_v & \xrightarrow{\quad g \quad} & a\,F\,h_v \\
{\scriptstyle \beta^u_{a\,F\,h_v}} \uparrow & & \uparrow {\scriptstyle \beta^u_{a\,F\,h_v}} \\
a\,F\,h_u & \xrightarrow[\quad g|_u \quad]{} & a\,F\,h_u
\end{array}$$

shows that $g|_u$ corresponds universally to g under the bijection

$$(a\,F\,h_u,\ a\,F\,h_v) \cong (a\,F\,h_u,\ a\,F\,h_u).$$ ■

Lemma 8.

Given $u \leqslant v$ *in* $\mathbb{H}$ *and* $f : a\,F\,h_u \to a\,F\,h_v$ *in* $\mathrm{Sh}(\mathbb{H}, \mathbb{T})$, *there is some operation* $\alpha \in \mathcal{O}_1(u)$ *such that* f *factors into the following composite :*

$$a\,F\,h_u \xrightarrow{\ \alpha\ } a\,F\,h_u \xhookrightarrow{\ \beta^u_{a\,F\,h_v}\ } a\,F\,h_v.$$

By the adjunction $u_! \dashv u^*$, f factors through $\beta^u_{a\,F\,h_v}$ into a morphism $\alpha : a\,F\,h_u \to a\,F\,h_u$. By lemma 7, α comes from some operation $\alpha \in \mathcal{O}_1(u)$. ■

Lemma 9.

Consider $u \not\leqslant v$ *in* $\mathbb{H}$ *and* $g : F\,h_u \to F\,h_v$ *in* $\mathrm{Pr}(\mathbb{H}, \mathbb{T})$. *Then for any subobject* $R \rightarrowtail F\,h_v$ *in* $\mathrm{Pr}(\mathbb{H}, \mathbb{T})$, $g^{-1}(R) = F\,h_u$ *and thus for any subobject* $S \rightarrowtail a\,F\,h_v$ *in* $\mathrm{Sh}(\mathbb{H}, \mathbb{T})$, $(ag)^{-1}(S) = a\,F\,h_u$.

We have $F\,h_u(u) = F1$ and $F\,h_v(u) = F0$. This proves that g_u takes the generic element of $F\,h_u(u)$ into a constant of $F\,h_v(u)$. Now if $w \leqslant u$, the generic element of $F\,h_u(w) = F1$ is the restriction of the generic element of $F\,h_u(u)$; it is thus sended to a constant in $F\,h_v(w)$. Thus for $w \leqslant u$, g_w takes values in the constants of $F\,h_v(w)$. On the other hand if $w \not\leqslant u$, $F\,h_u(w) = F0$ and g_w is easily seen to take its values in the constants of $F\,h_v(w)$. Thus g factors through 0. Now consider the following diagram where the square is a pullback

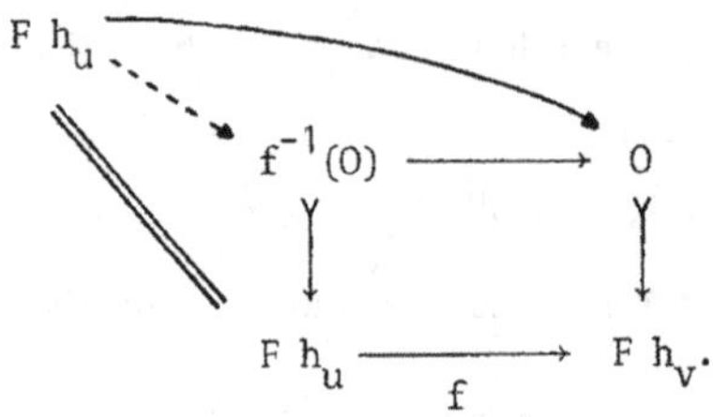

It shows that the monomorphism

$$f^{-1}(0) \rightarrowtail F\,h_u$$

is a regular epimorphism and thus an isomorphism. But

$$f^{-1}(R) \geqslant f^{-1}(0) = F\,h_u$$

and so $f^{-1}(R) = F\,h_u$.

Now if $S \rightarrowtail a\,F\,h_v$ is some subobject in $Sh(\mathbb{H}, \mathbb{T})$, $g^{-1}(S) = F\,h_u$ in $Pr(\mathbb{H}, \mathbb{T})$ and thus $(ag)^{-1}(S) = a\,F\,h_u$ in $Sh(\mathbb{H}, \mathbb{T})$. ■

The reader will point out that lemma 9 cannot be transposed to $Sh(\mathbb{H}, \mathbb{T})$. In $Sh(\mathbb{H}, \mathbb{T})$, the best approximation to the previous result is given by :

<u>Lemma 10</u>.

Let $u \not\leqslant v$ *in* $\mathbb{H}$ *and* $f : a\,F\,h_u \to a\,F\,h_v$ *in* $Sh(\mathbb{H}, \mathbb{T})$. *There exists some* $u_i \leqslant u$, $u_i \not\leqslant v$ *such that, for any subobject* $A \rightarrowtail a\,F\,h_v$ *in* $Sh(\mathbb{H}, \mathbb{T})$, $f^{-1}(A) \geqslant a\,F\,h_{u_i}$.

f corresponds by proposition I - 5 to some element x in $a\,F\,h_v(u)$. Thus there is a covering $u = \bigvee_{i \in I} u_i$ in $\mathbb{H}$ such that $x|_{u_i} \in F\,h_v(u_i)$. This produces morphisms $f_i : F\,h_{u_i} \to F\,h_v$ such that $a\,f_i$ is just the composite

$$a\,F\,h_{u_i} \xrightarrow{\beta^{u_i}} a\,F\,h_u \xrightarrow{\ f\ } a\,F\,h_v.$$

But $u = \bigvee_{i\in I} u_i \not\leqslant v$; so there is some i with $u_i \not\leqslant v$. For this i, by lemma 9 and the exactness of a :

$$f^{-1}(A) \geqslant (a\,f_i)^{-1}(A) = a\,F\,h_{u_i}.$$

■

Lemma 11.

Let u *be an element in* $\mathbb{H}$, A *and* B *objects in* $Sh(\mathbb{H},\mathbb{T})$. *Any morphisms* $f : A|_u \to B$ *in* $Sh(\mathbb{H},\mathbb{T})$ *factors through* $B|_u$. *In particular for any subobject* $R \rightarrowtail B$, $f^{-1}(R) = f^{-1}(R|_u)$.

By the adjunction $u_! \dashv u^*$, any morphism $f : u_!\, u^*A \to B$ factors through $u_!\, u^*B$. Therefore

$$f^{-1}(R) = f^{-1}(R \cap B|_u) = f^{-1}(R|_u)$$

by proposition II - 8. ■

§ 2. THE CANONICAL TOPOS $\mathcal{E}(\mathbb{H},\mathbb{T})$

In this paragraph, we consider again a commutative theory $\mathbb{T}$ in the topos $Sh(\mathbb{H})$ of sheaves on the frame $\mathbb{H}$. We define a monoid $M_{\mathbb{T}}$ in $Sh(\mathbb{H})$ and we consider the topos (cfr. [12]) $\mathcal{E}(\mathbb{H},\mathbb{T})$ of $M_{\mathbb{T}}$-objects in $Sh(\mathbb{H})$. In fact $M_{\mathbb{T}}$ turns out to be just the monoid $\mathcal{O}_1$ of 1-ary operations. This depends heavily on the peculiarity that the site $\mathbb{H}$ has at most one arrow between two objects. For a general site S, a description of the monoïd, similar to the one in definition 12 below, could still be given. But the monoïd so obtained would no longer be isomorphic to $\mathcal{O}_1$. But looking through succeeding constructions and proofs, one notices that the useful thing is definition 12 and not the monoïd $\mathcal{O}_1$.

Definition 12.

The monoïd $M_{\mathbb{T}}$ *in* $Sh(\mathbb{H})$ *is defined by*

$$M_{\mathbb{T}}(u) = (a\,F\,h_u,\ a\,F\,h_u)$$

with the usual composition of morphisms in $Sh(\mathbb{H},\mathbb{T})$ *and*

$$M_{\mathbb{T}}(u \leqslant v)(f) = f|_u.$$

By lemma 7, $M_{\mathbb{T}}$ is isomorphic to $\mathcal{O}_1$ and is thus an object in Sh($\mathbf{H}$); so definition 12 makes sense.

Definition 13.

The topos $\mathcal{E}(\mathbf{H}, \mathbb{T})$ *is the topos of* $M_{\mathbb{T}}$*-objects in* Sh($\mathbf{H}$).

Proposition 14.

Any object A *in* Sh($\mathbf{H}, \mathbb{T}$) *is canonically equipped with the structure of a* $M_{\mathbb{T}}$*-object. This gives rise to a forgetful functor* $U : \mathrm{Sh}(\mathbf{H}, \mathbb{T}) \to \mathcal{E}(\mathbf{H}, \mathbb{T})$. U *has a left adjoint.*

For any object A in Sh($\mathbf{H}, \mathbb{T}$) and any u in $\mathbf{H}$ the action is given by composition

$$\begin{array}{ccc} M_{\mathbb{T}}(u) \times A(u) & \cong & (a\,F\,h_u,\ a\,F\,h_1) \times (a\,F\,h_u,\ A) \\ & & \downarrow \\ A(u) & \cong & (a\,F\,h_u,\ A) \end{array}$$

via the isomorphisms of proposition I - 5. The action being defined by composition, it is obvious that A is made into a $M_{\mathbb{T}}$-object and moreover any morphism $f : A \to B$ in Sh($\mathbf{H}, \mathbb{T}$) commutes with the actions on A and B.

U is algebraic (we forget the n-ary operations of $\mathbb{T}$ for $n \neq 1$); thus it has a left adjoint. ■

Generally we shall use the same notation A for an object in Sh($\mathbf{H}, \mathbb{T}$) and its underlying object in $\mathcal{E}(\mathbf{H}, \mathbb{T})$.

§ 3. THE CLASSIFYING OBJECT $\Omega_{\mathbb{T}}$ FOR ALGEBRAIC SHEAVES

The topos $\mathcal{E}(\mathbf{H}, \mathbb{T})$ has a classifying object, but it plays no role in our investigation. So we can safely denote by $\Omega_{\mathbb{T}}$ the object in $\mathcal{E}(\mathbf{H}, \mathbb{T})$ which possesses classification properties with respect to Sh($\mathbf{H}, \mathbb{T}$).

Definition 15.

$\Omega_{\mathbb{T}}$ *is the presheaf on* $\mathbf{H}$ *whose value at* $u \in \mathbf{H}$ *is the set of subobjects of* $a\,F\,h_u$ *in* Sh($\mathbf{H}, \mathbb{T}$). *The restriction morphisms* $\Omega_{\mathbb{T}}(u \leq v)$ *act by restriction at* u : $\Omega_{\mathbb{T}}(u \leq v)(R) = R|_u$.

Proposition 16.

Ω_{II} is a sheaf on H.

Suppose given $u = \bigvee_{i \in I} u_i$ in H and R, S subobjects of a F h_u such that for any $i \in I$, $R|_{u_i} = S|_{u_i}$. By lemma 2,

$$R = \bigcup_{i \in I} R|_{u_i} = \bigcup_{i \in I} S|_{u_i} = S.$$

So Ω_{II} is a separated presheaf.

Suppose again we are given $u = \bigvee_{i \in I} u_i$ in H and for each i a subobject R_i of a F h_{u_i} such that

$$R_i|_{u_i \wedge u_j} = R_j|_{u_i \wedge u_j}.$$

Consider each R_i as a subobject of a F h_u. Take $R = \bigcup_{i \in I} R_i$. For any index j in J

$$
\begin{aligned}
R|_{u_j} &= a\,F\,h_{u_j} \cap R && \text{lemma 3}\\
&= a\,F\,h_{u_j} \cap \Big(\bigcup_{i \in I} R_i\Big) && \text{proposition II - 5}\\
&= \bigcup_{i \in I} (a\,F\,h_{u_j} \cap R_i) && \text{proposition II - 5}\\
&= \bigcup_{i \in I} (a\,F\,h_{u_j} \cap a\,F\,h_{u_i} \cap R_i) && R_i \subseteq a\,F\,h_{u_i}\\
&= \bigcup_{i \in I} R_i|_{u_i \wedge u_j} && \text{lemma 3}\\
&= \bigcup_{i \in I} R_j|_{u_i \wedge u_j}\\
&= R_j
\end{aligned}
$$

because one of the terms of this last union is

$$R_j|_{u_j \wedge u_j} = R_j.$$

Thus Ω_{II} is a sheaf. ■

Proposition 17.

$\Omega_{\mathbb{T}}$ *is canonically provided with the structure of a* $M_{\mathbb{T}}$*-object.*

For any u in $\mathbb{H}$ we define an action

$$M_{\mathbb{T}}(u) \times \Omega_{\mathbb{T}}(u) \cong (a\,F\,h_u,\ a\,F\,h_u) \times \Omega_{\mathbb{T}}(u) \to \Omega_{\mathbb{T}}(u)$$

by sending the pair (f, R) to $f^{-1}(R)$. So $\Omega_{\mathbb{T}}$ becomes a $M_{\mathbb{T}}$- object since $(fg)^{-1}(R) = g^{-1}(f^{-1}(R))$ and $id^{-1}(R) = R$. ∎

Proposition 18.

$\Omega_{\mathbb{T}}$ *has the structure of an inf-semi-lattice with greatest element in the topos* $\mathcal{E}(\mathbb{H}, \mathbb{T})$.

For any u in $\mathbb{H}$, $\Omega_{\mathbb{T}}(u)$ is the set of subobjects of $a\,F\,h_u$ in $Sh(\mathbb{H}, \mathbb{T})$: it is thus a complete lattice. In order to prove the proposition, we must show that the morphisms

$$\Omega_{\mathbb{T}}(u) \times \Omega_{\mathbb{T}}(u) \xrightarrow{\wedge_{\mathbb{T}}(u)} \Omega_{\mathbb{T}}(u);\ (R, S) \mapsto R \cap S$$

$$1 \xrightarrow{t_{\mathbb{T}}(u)} \Omega_{\mathbb{T}}(u) \qquad ;\ * \mapsto a\,F\,h_u$$

are the components of morphisms

$$\Omega_{\mathbb{T}} \times \Omega_{\mathbb{T}} \xrightarrow{\wedge_{\mathbb{T}}} \Omega_{\mathbb{T}}$$

$$1 \xrightarrow{t_{\mathbb{T}}} \Omega_{\mathbb{T}}$$

in the topos $\mathcal{E}(\mathbb{H}, \mathbb{T})$.

First we will prove that $\wedge_{\mathbb{T}}$ and $t_{\mathbb{T}}$ are morphisms in $Sh(\mathbb{H}, \mathbb{T})$. Given $u \leqslant v$ in $\mathbb{H}$ and R, S elements in $\Omega_{\mathbb{T}}(u)$, we have to verify that

$$R|_u \cap S|_u = (R \cap S)|_u$$

$$a\,F\,h_v|_u = a\,F\,h_u.$$

The first equality holds because each side equals $R \cap S \cap a\,F\,h_u$ (lemma 3) and the second equality holds by theorem I - 5.

To have morphisms in $\mathcal{E}(\mathbb{H}, \mathbb{T})$ we must prove moreover that, for

$\alpha : a\,F\,h_u \to a\,F\,h_u$ in $Sh(\mathbb{H}, \mathbb{T})$,

$$\alpha^{-1}(R \cap S) = \alpha^{-1}(R) \cap \alpha^{-1}(R)$$

$$\alpha^{-1}(a\,F\,h_u) = a\,F\,h_u$$

which is obvious. ■

The reader will point out that, even if each $\Omega_{\mathbb{T}}(u)$ is a complete lattice, $\Omega_{\mathbb{T}}$ is generally not a complete lattice in $\mathcal{E}(\mathbb{H}, \mathbb{T})$. Indeed, for $\alpha : a\,F\,h_u \to a\,F\,h_u$ in $Sh(\mathbb{H}, \mathbb{T})$ and R_i in $\Omega_{\mathbb{T}}(u)$, we have in general

$$\alpha^{-1}\Big(\bigcup_{i \in I} R_i\Big) \neq \bigcup_{i \in I} \alpha^{-1}(R_i)$$

$$\alpha^{-1}(0) \neq 0.$$

Thus suprema and smallest element do not exist for $\Omega_{\mathbb{T}}$ in $\mathcal{E}(\mathbb{H}, \mathbb{T})$. On the other hand it is possible, using the classifying object Ω of the topos $\mathcal{E}(\mathbb{H}, \mathbb{T})$, to define arbitrary infima for $\Omega_{\mathbb{T}}$ in $\mathcal{E}(\mathbb{H}, \mathbb{T})$, but we will not consider this question here.

§ 4. CLASSIFICATION OF SUBOBJECTS IN $Sh(\mathbb{H}, \mathbb{T})$

We shall use now the topos $\mathcal{E}(\mathbb{H}, \mathbb{T})$ and the object $\Omega_{\mathbb{T}}$ to classify the subobjects $R \rightarrowtail A$ in $Sh(\mathbb{H}, \mathbb{T})$ by morphisms $\varphi_R : A \to \Omega_{\mathbb{T}}$ in $\mathcal{E}(\mathbb{H}, \mathbb{T})$. Such a morphism φ_R will be called the characteristic map of the subobject R of A.
But not all morphisms in $\mathcal{E}(\mathbb{H}, \mathbb{T})$ with codomain $\Omega_{\mathbb{T}}$ will be suitable. We do need an additional feature :

Definition 19.

Let A be any object in $Sh(\mathbb{H}, \mathbb{T})$.
A characteristic map φ on A is a morphism $\varphi : A \to \Omega_{\mathbb{T}}$ in $\mathcal{E}(\mathbb{H}, \mathbb{T})$ such that for any $n \in \mathbb{N}$, $u \in \mathbb{H}$ and $\beta \in O_n(u)$

$$(\wedge_{\mathbb{T}} \circ \varphi^n)\big|_u \leq \varphi\big|_u \circ \beta.$$

The latter inequality corresponds to the following diagram :

$$\begin{array}{ccc} A^n\big|_u & \xrightarrow{\ \beta\ } & A\big|_u \\ {\scriptstyle \varphi^n|_u}\Big\downarrow & \leqslant & \Big\downarrow{\scriptstyle \varphi|_u} \\ \Omega^n_{\mathbb{T}}\big|_u & \xrightarrow[\ \wedge^n_{\mathbb{T}}|_u\ ]{} & \Omega_{\mathbb{T}}\big|_u \end{array}$$

where β exists by lemma 6 applied to Sh(u↓, $\mathbb{T}_{u\downarrow}$). In other words, definition 19 says that φ is a characteristic map as soon as for any $\beta \in \mathcal{O}^n(u)$, $v \leqslant u$ and $x_1, \ldots, x_n \in A(v)$

$$\bigwedge_{i=1}^{n} \varphi_v(x_i) \leqslant \varphi_v(\beta(x_1, \ldots, x_n)).$$

This notion of characteristic map will be a useful tool when classifying the localizations of Sh(H, $\mathbb{T}$). The reason for the name "characteristic map" will be found in theorem 22.

Some more remarks about this definition. If n = 1, clearly we use the convention $\wedge^n_{\mathbb{T}} = id_{\Omega_{\mathbb{T}}}$ and if n = 0, $\wedge^0_{\mathbb{T}} = t_{\mathbb{T}}$. We know (lemma 5) that for $n \geqslant 1$, $A^n\big|_u$ and $(A\big|_u)^n$ coincide, so there cannot be any confusion in the notations in this case; in particular $\varphi^n\big|_u = (\varphi\big|_u)^n$ acts componentwise. Now if n = 0, the diagram reduces to

$$\begin{array}{ccc} 1\big|_u & \xrightarrow{\ \beta\ } & A\big|_u \\ \Big\| & \leqslant & \Big\downarrow{\scriptstyle \varphi|_u} \\ 1\big|_u & \xrightarrow[\ t_{\mathbb{T}}|_u\ ]{} & \Omega_{\mathbb{T}}\big|_u \end{array}$$

which is thus equivalent to the commutativity and again no confusion is possible.

If we think of φ as the map which characterizes those elements x in A(u) such that $\varphi_u(x) = 1$, we deduce from definition 19 that

$$\varphi_n(x_1) = \ldots = \varphi_n(x_n) = 1 \Rightarrow \varphi_n(\beta(x_1, \ldots, x_n)) = 1.$$

This formula thus expresses the fact that those x are stable under the operations of $\mathbb{T}$ (= are a subobject in Sh(H, $\mathbb{T}$)). But definition 19 conveys more than this simple fact :

Proposition 20.

Let f : A → B *be a morphism in* Sh(H, T) *and* φ : B → $\Omega_{\mathbb{T}}$ *a characteristic map in* &(H, T). *Then* φ ∘ f *is again a characteristic map.*

Consider the following diagram for any $n \in \mathbb{N}$, $u \in H$ and $\beta \in \mathcal{O}_n(u)$:

$$\begin{array}{ccc}
A^n|_u & \xrightarrow{\quad\beta\quad} & A|_u \\
{\scriptstyle f^n|_u}\downarrow & \hookrightarrow & \downarrow{\scriptstyle f|_u} \\
B^n|_u & \xrightarrow[\quad\beta\quad]{} & B|_u \\
{\scriptstyle \varphi^n|_u}\downarrow & \leqslant & \downarrow{\scriptstyle \varphi|_u} \\
\Omega^n_{\mathbb{T}}|_u & \xrightarrow[\wedge^n_{\mathbb{T}}|_u]{} & \Omega_{\mathbb{T}}|_u .
\end{array}$$

The first square is commutative because f is a morphism in Sh(H, T). Therefore φ ∘ f satisfies the conditions of definition 19. ∎

Proposition 21.

Let A *be an object in* Sh(H, T), φ : A → $\Omega_{\mathbb{T}}$ *a characteristic map,* u ∈ H *and* $\beta \in \mathcal{O}_0(u)$. *Then* $\varphi_u(\beta) = a\,F\,h_u$.

For we have seen that $\varphi|_u \circ \beta = t_{\mathbb{T}}|_u$. ∎

Theorem 22.

Let A *be some object in* Sh(H, T).

There is a bijection between

(1) *the subobjects* R ↣ A *in* Sh(H, T).

(2) *the characteristic maps* φ : A → $\Omega_{\mathbb{T}}$ *in* &(H, T).

Moreover, if R *and* φ *correspond to each other under this bijection, the following square is a pullback in the topos* &(H, T) :

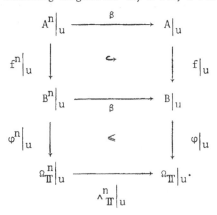

$$\begin{array}{ccc} R & \longrightarrow & 1 \\ \downarrow & & \downarrow t_{\mathbb{T}} \\ A & \xrightarrow[\varphi]{} & \Omega_{\mathbb{T}} \end{array}$$

Consider a subobject $R \rightarrowtail A$ in $Sh(\mathbb{H}, \mathbb{T})$. We shall define a characteristic map $\varphi : A \to \Omega_{\mathbb{T}}$ in $\mathcal{E}(\mathbb{H}, \mathbb{T})$. By proposition I - 5, for any u in $\mathbb{H}$ we define

$$\varphi(u) : A(u) \cong (a\ F\ h_u, A) \to \Omega_{\mathbb{T}}(u)$$
$$\varphi(u)(f) = f^{-1}(R).$$

φ is a natural transformation since for any g in $(a\ F\ h_v, A)$ with $u \leq v$ we have

$$\varphi(u)(g|_u) = g|_u^{-1}(R) = g^{-1}(R) \cap a\ F\ h_u = g^{-1}(R)|_u.$$

Moreover, each $\varphi(u)$ is a morphism in $\mathcal{E}(\mathbb{H}, \mathbb{T})$ since for any $\alpha \in (a\ F\ h_u, a\ F\ h_u)$ and $f \in (a\ F\ h_u, A)$,

$$\alpha^{-1}(f^{-1}(R)) = (f \circ \alpha)^{-1}(R).$$

So φ is a morphism in $\mathcal{E}(\mathbb{H}, \mathbb{T})$.

We prove now that φ is a characteristic map (notations of definition 19). If $n \in \mathbb{N}$, $\beta \in \mathcal{O}_n(u)$, $v \leq u$ and $(f_1, \ldots, f_n) \in A^n|_u(v) = A^n(v) = (a\ F\ h_v, A)^n$, we need to show that

$$\beta(f_1, \ldots, f_n)^{-1}(R) \geq \bigcap_{i=1}^{n} f_i^{-1}(R),$$

which means, by lemma 11

$$\beta(f_1, \ldots, f_n)^{-1}(R|_v) \geq \bigcap_{i=1}^{n} f_i^{-1}(R|_v).$$

This is true by lemma 4 applied to $Sh(u\!\downarrow, \mathbb{T}_u)$. (By lemma 5, it does not matter where inverse images are computed).

The definition of φ shows that $f \in A(u)$ is sended to $t_{\mathbb{T}}(*)$ in $\Omega_{\mathbb{T}}(u)$ if and only if $f^{-1}(R) = a\ F\ h_u$, i.e. if and only if f factors through R :

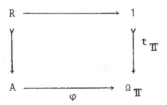

$$\begin{array}{ccc} f^{-1}(R) & \longrightarrow & R \\ \| & \text{p.b.} & \downarrow \\ a\,F\,h_u & \underset{f}{\longrightarrow} & A. \end{array}$$

By proposition I - 5, this shows exactly that

$$\varphi_u^{-1}(1) = (a\,F\,h_u,\ R) = R(u).$$

In other words, the following square is a pullback in $\mathcal{E}(\mathbb{H}, \mathbb{T})$

$$\begin{array}{ccc} R & \longrightarrow & 1 \\ \downarrow & \text{p.b.} & \downarrow t_{\mathbb{T}} \\ A & \underset{\varphi}{\longrightarrow} & \Omega_{\mathbb{T}}. \end{array}$$

Conversely, consider a characteristic map $\varphi : A \to \Omega_{\mathbb{T}}$ in $\mathcal{E}(\mathbb{H}, \mathbb{T})$, with A in $Sh(\mathbb{H}, \mathbb{T})$. Define $R \rightarrowtail A$ by the following pullback in $\mathcal{E}(\mathbb{H}, \mathbb{T})$

$$\begin{array}{ccc} R & \longrightarrow & 1 \\ \downarrow & \text{p.b.} & \downarrow t_{\mathbb{T}} \\ A & \underset{\varphi}{\longrightarrow} & \Omega_{\mathbb{T}}. \end{array}$$

It remains to show that R is a subobject of A in $Sh(\mathbb{H}, \mathbb{T})$, i.e. is stable under the operations of $\mathbb{T}$. Consider $u \in \mathbb{H}$, $n \in \mathbb{N}$, $\beta \in \mathcal{O}_n(u)$ and $x_1, \ldots, x_n \in R(u)$. Thus $\varphi_u(x_i) = a\,F\,h_u$.

$$\varphi_u(\beta(x_1, \ldots, x_n)) \geqslant \bigwedge_{i=1}^{n} \varphi_u(x_i) \qquad \text{definition 19}$$

$$= \bigwedge_{i=1}^{n} a\,F\,h_u$$

$$= a\,F\,h_u$$

and this shows that $\beta(x_1, \ldots, x_n)$ is in $R(u)$.

We must prove that we have defined a one to one correspondance between the subobjects of A in $Sh(\mathbb{H}, \mathbb{T})$ and the characteristic maps on A in $\mathcal{E}(\mathbb{H}, \mathbb{T})$. If we start with a subobject $R \rightarrowtail A$ in $Sh(\mathbb{H}, \mathbb{T})$, we construct a map $\varphi : A \to \Omega_{\mathbb{T}}$ in $\mathcal{E}(\mathbb{H}, \mathbb{T})$ and finally the subobject $\varphi^{-1}(t_{\mathbb{T}}) \rightarrowtail A$ in $Sh(\mathbb{H}, \mathbb{T})$. But this subobject is just R, as follows from the pullback

$$\begin{array}{ccc} R & \longrightarrow & 1 \\ \downarrow & \text{p.b.} & \downarrow t_{\mathbb{T}} \\ A & \xrightarrow[\varphi]{} & \Omega_{\mathbb{T}} \end{array}$$

Conversely consider a characteristic map $\varphi : A \to \Omega_{\mathbb{T}}$ in $\mathcal{E}(\mathbb{H},\mathbb{T})$, the corresponding subobject $R = \varphi^{-1}(t_{\mathbb{T}})$ of A in $Sh(\mathbb{H},\mathbb{T})$ and its characteristic map $\varphi_R : A \to \Omega_{\mathbb{T}}$ in $\mathcal{E}(\mathbb{H},\mathbb{T})$. We must prove the equality $\varphi = \varphi_R$. Thus if $u \in \mathbb{H}$ and $f \in (a\,F\,h_u, A) \cong A(u)$ we have to show that $\varphi(u)(f) = \varphi_R(u)(f)$. But $\varphi_R(u)(f) = f^{-1}(R)$ by definition of φ_R and thus we need to prove that $\varphi(u)(f) = f^{-1}(R)$. But $\varphi(u)(f) \in \Omega_{\mathbb{T}}(u)$ is some subobject of $a\,F\,h_u$: call it S. We shall prove that $S = (\varphi \circ f)^{-1}(t_{\mathbb{T}})$. Therefore

$$\begin{aligned} \varphi(u)(f) &= S \\ &= f^{-1}(\varphi^{-1}(t_{\mathbb{T}})) \\ &= f^{-1}(R) \end{aligned}$$

and the proof will be complete.

So we must verify that the following square is a pullback in $\mathcal{E}(\mathbb{H},\mathbb{T})$:

$$\begin{array}{ccccc} S & & \longrightarrow & & 1 \\ \downarrow & & & & \downarrow t_{\mathbb{T}} \\ a\,F\,h_u & \xrightarrow[f]{} & A & \xrightarrow[\varphi]{} & \Omega_{\mathbb{T}} \end{array}$$

Thus for any $v \in \mathbb{H}$ and $g \in (a\,F\,h_v, a\,F\,h_u) \cong a\,F\,h_u(v)$ we must prove that

$$(\varphi\ f)(v)(g) = a\,F\,h_v \quad \text{iff} \quad g \in (a\,F\,h_v, S) \cong S(v).$$

But $g \in a\,F\,h_u(v)$ and by construction of the associated sheaf functor, there is a covering $(v_i)_{i\in I}$ of v ($v = \bigvee_{i\in I} v_i$) such that for any i, $g|_{v_i} = a(g_i)$ where $g_i \in F\,h_u(v_i)$ or, equivalently, $g_i : F\,h_{v_i} \to F\,h_u$ is a morphism in $Pr(\mathbb{H},\mathbb{T})$.
By lemma 2, the statement above reduces to :

$$\forall\, i \in I \quad (\varphi\ f)(v)(g)\big|_{v_i} = a\,F\,h_v\big|_{v_i} \quad \text{iff} \quad g\big|_{v_i} \in S(v_i)$$

or in other words

$$\forall\, i \in I \quad (\varphi\ f)(v_i)(ag_i) = a\,F\,h_{v_i} \quad \text{iff} \quad ag_i \in (a\,F\,h_{v_i}, S).$$

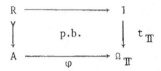

Suppose first that $v_i \leqslant u$. By lemma 8, ag_i factors through

$$\begin{array}{ccc} a\,F\,h_{v_i} & \xrightarrow{\quad ag_i \quad} & a\,F\,h_u \\ & \searrow^{\alpha} \quad \hookrightarrow \quad \nearrow_{\beta^{v_i}_{a\,F\,h_u}} & \\ & a\,F\,h_{v_i} & \end{array}$$

From this we deduce

$$\begin{aligned}
&(\varphi\, f)(v_i)(ag_i) \\
&\quad = (\varphi\, f)(v_i)(\beta^{v_i} \circ \alpha) \\
&\quad = \alpha^{-1}((\varphi\, f)(v_i)(\beta^{v_i})) && \varphi\, f \in \mathcal{E}(\mathbb{H}, \mathbb{T}) \\
&\quad = \alpha^{-1}((\varphi\, f)(v_i)(id_{a\,F\,h_u}|_{v_i})) \\
&\quad = \alpha^{-1}((\varphi\, f)(u)(id_{a\,F\,h_u}))|_{v_i} && \varphi\, f \in Sh(\mathbb{H}) \\
&\quad = \alpha^{-1}(S|_{v_i}) && \text{Yoneda lemma} \\
&\quad = \alpha^{-1}(\beta^{v_i})^{-1}(S) \\
&\quad = (ag_i)^{-1}(S).
\end{aligned}$$

Therefore

$$\begin{aligned}
(\varphi\, f)(v_i)(ag_i) &= a\,F\,h_{v_i} \\
&\text{iff } (ag_i)^{-1}(S) = a\,F\,h_{v_i} \\
&\text{iff } ag_i \in (a\,F\,h_{v_i}, S)
\end{aligned}$$

and this concludes the proof with $v_i \leqslant u$.

Now suppose $v_i \not\leqslant u$. Then $g_i \in F\,h_u(v_i) = F0 \subseteq S(v_i)$. On the other hand, propositions 20 and 21 imply that $(\varphi\, f)(v_i)(ag_i) = a\,F\,h_{v_i}$. Hence, the result is trivial. ■

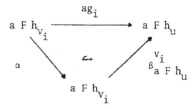

§ 5. UNIVERSAL CLOSURE OPERATIONS ON Sh(H, T)

As before, suppose the theroy T to be commutative. We now proceed to investigate the localizations of Sh(H, T), i.e. the full reflexive subcategories of Sh(H, T) whose reflection is an exact functor. If T is the initial theory, Sh(H, T) is the topos Sh(H) and its localizations can be classified at the same time by Grothendieck topologies on H, by Lawvere - Tierney topologies on the Ω-object of Sh(H) and by universal closure operations on Sh(H). This chapter is devoted to generalize these results to Sh(H, T). In the present paragraph, we characterize a localization of Sh(H, T) by a universal closure operation.

Definition 23.

A universal closure operation on Sh(H, T) *is defined by specifying for each subobject* $R \rightarrowtail A$ *in* Sh(H, T), *a subobject* $\overline{R} \rightarrowtail A$ *in* Sh(H, T) *such that, for any subobjects* R, S *of* A *and any morphism* $f : B \to A$

(c 1) $R \leqslant \overline{R}$

(c 2) $R \leqslant S \Rightarrow \overline{R} \leqslant \overline{S}$

(c 3) $\overline{\overline{R}} = \overline{R}$

(c 4) $\overline{f^{-1}(R)} = f^{-1}(\overline{R})$.

Proposition 24.

An equivalent system of axioms for a closure operation on Sh(H, T) *is given by (with the notations of definition* 23*)*

(c 5) $\overline{A} = A$

(c 6) $\overline{R \cap S} = \overline{R} \cap \overline{S}$

(c 3) $\overline{\overline{R}} = \overline{R}$

(c 4) $\overline{f^{-1}(R)} = f^{-1}(\overline{R})$.

Start from (c 1) to (c 4). By (c 1), $A \leqslant \overline{A}$ and thus $\overline{A} = A$; this is (c 5). Now $R \cap S \leqslant R$ and thus, by (c 2), $\overline{R \cap S} \leqslant \overline{R}$; in the same way $\overline{R \cap S} \leqslant \overline{S}$ and finally $\overline{R \cap S} \leqslant \overline{R} \cap \overline{S}$. Now by (c 1) the following square is a pullback

$$\begin{array}{ccc} R \cap S & \hookrightarrow & S \\ \downarrow & & \downarrow \\ R \cap \overline{S} & \hookrightarrow & \overline{S}. \end{array}$$

This shows that the closure of $R \cap S$ in $R \cap \overline{S}$ is $R \cap \overline{S}$ (by (c 4)); but this closure is also $(\overline{R \cap S}) \cap (R \cap \overline{S})$ (again by (c 4)). Thus

$$R \cap \overline{S} = (\overline{R \cap S}) \cap (R \cap \overline{S})$$

which shows that $R \cap \overline{S} \leqslant \overline{R \cap S}$. Therefore, by (c 2) - (c 3) and this last relation applied to R, S and $\overline{S}$, R we obtain :

$$\overline{R \cap S} = \overline{\overline{R \cap S}} \geqslant \overline{R \cap \overline{S}} \geqslant \overline{R} \cap \overline{S}.$$

This implies (c 6).

Conversely start from (c 3) to (c 6). By (c 4), intersect R, $\overline{R}$ with R : $R \cap \overline{R}$ is the closure of $R \cap R$ in $R \cap R$, i.e. $R \cap \overline{R} = R$ by (c 5). This proves that $R \leqslant \overline{R}$, which is (c 1). On the other hand

$$\begin{aligned} R \leqslant S \quad & \Rightarrow \quad R \cap S = R \\ & \Rightarrow \quad \overline{R} \cap \overline{S} = \overline{R} \qquad \text{by (c 6)} \\ & \Rightarrow \quad \overline{R} \leqslant \overline{S} \end{aligned}$$

which is (c 2). ■

<u>Proposition 25.</u>

Let $\mathbb{C} \underset{i}{\overset{\ell}{\rightleftarrows}} Sh(\mathbb{H}, \mathbb{T})$ *be a localization of* Sh($\mathbb{H}$, $\mathbb{T}$). *For any subobject* $R \overset{r}{\rightarrowtail} A$ *in* Sh($\mathbb{H}$, $\mathbb{T}$) *define* $\overline{R}$ *by the following pullback, where* η_A *is the universal morphism of the adjunction* $\ell \dashv i$,

$$\begin{array}{ccc} \overline{R} & \longrightarrow & i\ell(R) \\ {\scriptstyle \overline{r}} \big\downarrow & \text{p.b.} & \big\downarrow {\scriptstyle i\ell(r)} \\ A & \underset{\eta_A}{\longrightarrow} & i\ell(A) \end{array}$$

This is a universal closure operation on Sh($\mathbb{H}$, $\mathbb{T}$).

First point out that by exactness of ℓ, $\ell(r)$ and thus $\overline{r}$ are monomorphisms.

From the following commutative diagram

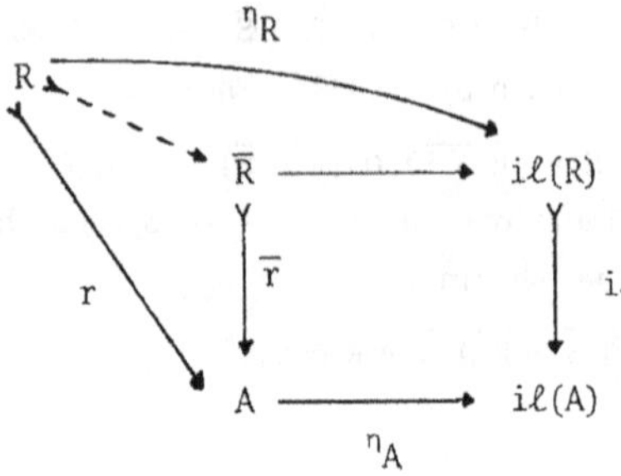

we deduce that $R \leqslant \bar{R}$, which is (c 1).

Now $R \leqslant S$ implies $i\ell(R) \leqslant i\ell(S)$ and thus $\eta_A^{-1}(i\ell\ R) \leqslant \eta_A^{-1}(i\ell\ S)$; this is (c 2).

Now i is full and faithful, thus $i\ell i\ell$ is isomorphic to $i\ell$ and $i\ell(\eta_A)$ is an isomorphism. Let us apply $i\ell$ to the diagram defining $\bar{R}$; ℓ beeing exact the new diagram is again a pullback :

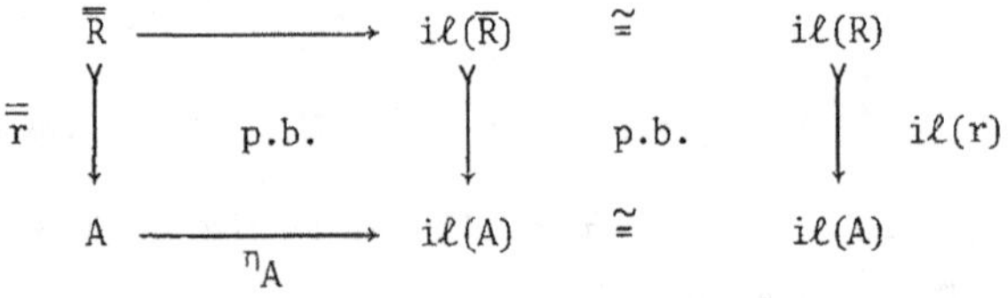

The composition of these pullbacks shows that $\bar{\bar{R}} = \bar{R}$, which is (c 3).

Finally if $f : B \to A$ is any morphism, the following cube is commutative

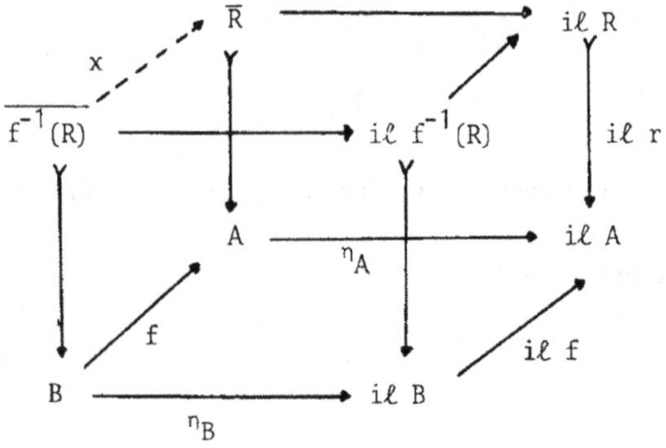

where x is the unique factorization through the pullback $\overline{R}$. The faces over η_A and η_B are pullbacks; the face over $i\ell$ f is a pullback by exactness of ℓ. By [21] - 7 - 8 - 4, the face over f is also a pullback. This proves that $f^{-1}(\overline{R}) = \overline{f^{-1}(R)}$, which is (c 4). ■

§ 6. LAWVERE - TIERNEY $\mathbb{T}$-TOPOLOGIES ON $\mathbb{H}$

Let $\mathbb{H}$ be a frame. A localization of Sh($\mathbb{H}$) (i.e. a subtopos of Sh($\mathbb{H}$)) can be classified by a Lawvere - Tierney topology $j : \Omega \to \Omega$ on the Ω-object of the topos Sh($\mathbb{H}$). Now, if $\mathbb{T}$ is any commutative theory in Sh($\mathbb{H}$), a localization of Sh($\mathbb{H}, \mathbb{T}$) will in a similar way be classified by a morphism $j : \Omega_{\mathbb{T}} \to \Omega_{\mathbb{T}}$ in the topos $\mathcal{E}$($\mathbb{H}, \mathbb{T}$). This morphism j satisfies the three well-known conditions for a Lawvere - Tierney topologies. For this reason, we call it a Lawvere - Tierney $\mathbb{T}$-topology on $\mathbb{H}$. The present paragraph is devoted to the construction of such a j from a universal closure operation on Sh($\mathbb{H}, \mathbb{T}$).

Definition 26.

Let $\mathbb{H}$ be a frame and $\mathbb{T}$ a commutative theory in Sh($\mathbb{H}$). *A Lawvere - Tierney $\mathbb{T}$-topology on $\mathbb{H}$ is a morphism $j : \Omega_{\mathbb{T}} \to \Omega_{\mathbb{T}}$ in the topos $\mathcal{E}$($\mathbb{H}, \mathbb{T}$) such that*

(L T 1) $j \circ t_{\mathbb{T}} = t_{\mathbb{T}}$

(L T 2) $j \circ j = j$

(L T 3) $j \circ \wedge_{\mathbb{T}} = \wedge_{\mathbb{T}} \circ (j \times j)$.

Proposition 27.

Let $(R \mapsto \overline{R})$ *be a universal closure operation on* Sh($\mathbb{H}, \mathbb{T}$). *For any* u *in* $\mathbb{H}$, *define* $j(u) : \Omega_{\mathbb{T}}(u) \to \Omega_{\mathbb{T}}(u)$ *by* $j(u)(R) = \overline{R}$. *This defines a morphism* $j : \Omega_{\mathbb{T}} \to \Omega_{\mathbb{T}}$ *which is a Lawvere - Tierney* $\mathbb{T}$*-topology on* $\mathbb{H}$.

If $v \leqslant u$ in $\mathbb{H}$, we know by (c 4) that for any R in $\Omega_{\mathbb{T}}(u)$, $\overline{R|_v} = \overline{R}|_v$. This says exactly that j is a morphism in Sh($\mathbb{H}$). Now for any $f : a\, F\, h_u \to a\, F\, h_u$, (c 4) also implies that $\overline{f^{-1}(R)} = f^{-1}(\overline{R})$. The latter equality simply states that j is a morphism in the topos $\mathcal{E}$($\mathbb{H}, \mathbb{T}$). Now (L T 1 - 2 - 3) follow immediately from (c 5 - 3 - 6). ■

Proposition 28.

With j given as in proposition 27, the image $I \rightarrowtail \Omega_{\mathbb{T}}$ *of j is such that for any* u *in* $\mathbb{H}$

$$I(u) = \{R \in \Omega_{\mathbb{T}}(u) \mid R = \overline{R}\}.$$

By definition of j, the elements in I(u) have the form $R = \overline{S}$ with $S \in \Omega_{\mathbb{T}}(u)$; thus by (c 3), $\overline{R} = \overline{\overline{S}} = \overline{S} = R$. Conversely if $R = \overline{R}$, then $R = j_u(R)$ and R is in I(u). ∎

Proposition 29.

With j *given as in proposition* 27, *if* $R \rightarrowtail A$ *is a subobject in* Sh($\mathbb{H}, \mathbb{T}$) *with characteristic map* φ, *then* $\overline{R}$ *has characteristic map* $j \circ \varphi$.

Consider the following diagram, where the right hand square is commutative and $u \in \mathbb{H}$, $\beta \in \mathcal{O}_n(u)$

$$\begin{array}{ccccc}
A^n\big|_u & \xrightarrow{\ \varphi^n|_u\ } & \Omega^n_{\mathbb{T}}\big|_u & \xrightarrow{\ j^n|_u\ } & \Omega^n_{\mathbb{T}}\big|_u \\
\beta\downarrow & \geqslant & \downarrow \wedge^n_{\mathbb{T}}\big|_u & \hookrightarrow & \downarrow \wedge^n_{\mathbb{T}}\big|_u \\
A\big|_u & \xrightarrow[\ \varphi|_u\ ]{} & \Omega_{\mathbb{T}}\big|_u & \xrightarrow[\ j|_u\ ]{} & \Omega_{\mathbb{T}}\big|_u
\end{array}$$

It shows that

$$\wedge^n_{\mathbb{T}}\big|_u \circ (j \circ \varphi)^n\big|_u$$

$$= \wedge^n_{\mathbb{T}}\big|_u \circ j^n\big|_u \circ \varphi^n\big|_u$$

$$= j\big|_u \circ \wedge^n_{\mathbb{T}}\big|_u \circ \varphi^n\big|_u$$

$$\leqslant j\big|_u \circ \varphi\big|_u \circ \beta$$

$$= (j \circ \varphi)\big|_u \circ \beta$$

since j is order preserving (by (c 2)). So $j \circ \varphi$ is a characteristic map on A and the subobject of A classified by $j \circ \varphi$ is $(j \circ \varphi)^{-1}(t_{\mathbb{T}})$ - (cfr. theorem 22).

Now if $u \in \mathbb{H}$, by proposition I - 5 an element in A(u) is just a morphism

$f : a\,F\,h_u \to A$. But $\varphi(u)$ was defined by $\varphi(u)(f) = f^{-1}(R)$, (cfr. theorem 22). Therefore, by (c 4)

$$
\begin{aligned}
&(j \circ \varphi)^{-1}(t_{\mathbb{T}})(u) \\
&\quad = \{f : a\,F\,h_u \to A \mid (j \circ \varphi)(u)(f) = a\,F\,h_u\} \\
&\quad = \{f : a\,F\,h_u \to A \mid j(u)(f^{-1}(R)) = a\,F\,h_u\} \\
&\quad = \{f : a\,F\,h_u \to A \mid \overline{f^{-1}(R)} = a\,F\,h_u\} \\
&\quad = \{f : a\,F\,h_u \to A \mid f^{-1}(\overline{R}) = a\,F\,h_u\} \\
&\quad = \{f : a\,F\,h_u \to A \mid f \text{ factors through } \overline{R}\} \\
&\quad = \overline{R}(u).
\end{aligned}
$$

An this shows that $\overline{R} = (j \circ \varphi)^{-1}(t_{\mathbb{T}})$, which concludes the proof. ■

§ 7. GABRIEL - GROTHENDIECK $\mathbb{T}$-TOPOLOGIES ON $\mathbb{H}$

For the topos Sh($\mathbb{H}$), the equivalence between Lawvere - Tierney topologies and Grothendieck topologies is well-known. A Grothendieck topology on $\mathbb{H}$ is given by specifying certain subobjects of the representable functors h_u ($u \in \mathbb{H}$). Now if you want to investigate sheaves of abelian groups on $\mathbb{H}$, a Gabriel localizing system on $\mathbb{H}$ is defined by specifying certain additive subobjects of the additive representable functors $\tilde{h}_u$ ($u \in \mathbb{H}$) on the free additive category $\tilde{\mathbb{H}}$ generated by $\mathbb{H}$ (cfr. [6] or [20]). Then the Gabriel localizing systems on $\mathbb{H}$ classify exactly the localizations of the category of presheaves of abelian groups on $\mathbb{H}$. Now observe that the axioms for a Gabriel localizing system are exactly the additive version of the axioms for a Grothendieck topology. This brings us in this paragraph, to define the concept of Gabriel - Grothendieck $\mathbb{T}$-topology on $\mathbb{H}$, for a commutative theory $\mathbb{T}$ in Sh($\mathbb{H}$). This is done by specifying for each $a\,F\,h_u$, ($u \in \mathbb{H}$), in Sh($\mathbb{H}, \mathbb{T}$) a certain set of subobjects. These subobjects satisfy the three usual Gabriel - Grothendieck axioms plus an additional axiom which takes into consideration the fact that we are dealing with sheaves and not simply presheaves as in the Gabriel and Grothendieck cases.

Definition 30.

Let $\mathbb{H}$ *be a frame and* $\mathbb{T}$ *a commutative theory in* Sh($\mathbb{H}$). *A Gabriel - Grothendieck* $\mathbb{T}$*-topology on* $\mathbb{H}$ *is defined by specifying for each* u *in* $\mathbb{H}$, *a family* $J(u)$ *of subobjects of* $a\,F\,h_u$ *in* Sh($\mathbb{H}, \mathbb{T}$) *such that*

(G G 1) $a\,F\,h_u \in J(u)$

(G G 2) $(R \in J(u)$ *and* $f : a\,F\,h_v \to a\,F\,h_u) \Rightarrow (f^{-1}(R) \in J(v))$

(G G 3) $\begin{pmatrix} R \in J(u)\ ;\ S \rightarrowtail a\,F\,h_u \\ \forall v \in H \quad \forall f : a\,F\,h_v \to R \quad f^{-1}(R \cap S) \in J(v) \end{pmatrix} \Rightarrow (S \in J(u))$

(G G 4) $\begin{pmatrix} R \rightarrowtail a\,F\,h_u\ ;\ u = \bigvee_{i \in I} u_i \text{ in } H \\ \forall i \in I \quad R|_{u_i} \in J(u_i) \end{pmatrix} \Rightarrow (R \in J(u)).$

Proposition 31.

Let j *be a Lawvere - Tierney* $\mathbb{T}$*-topology on* $\mathbf{H}$*. For any* u *in* $\mathbf{H}$*, consider*

$$J(u) = \{R \in \Omega_{\mathbb{T}}(u) \mid j(u)(R) = a\,F\,h_u\}.$$

This defines a Gabriel - Grothendieck $\mathbb{T}$*-topology* J *on* $\mathbf{H}$.

By (L T 1), (G G 1) holds. Now consider a morphism $f : a\,F\,h_v \to a\,F\,h_u$ in Sh(H, $\mathbb{T}$). By lemma 8, f can be factored into

$$a\,F\,h_v \xrightarrow{\alpha} a\,F\,h_v \hookrightarrow a\,F\,h_u .$$

Therefore for any R in J(u)

$$f^{-1}(R) = \alpha^{-1}(R|_v).$$

But j is a morphism in Sh(H), thus

$$j(v)(R|_v) = j(u)(R)|_v = a\,F\,h_u|_v = a\,F\,h_v .$$

Moreover, j is a morphism in &(H, $\mathbb{T}$), thus

$$j(v)(\alpha^{-1}(R|_v)) = \alpha^{-1}(j(v)(R|_v)) = \alpha^{-1}(a\,F\,h_v) = a\,F\,h_v .$$

Finally we have shown that

$$j(v)(f^{-1}(R)) = a\,F\,h_v$$

which proves that R is in J(v). This implies (G G 2).

We consider now the assumptions of (G G 3) and we denote by $r : R \rightarrowtail a\,F\,h_u$ the inclusion of R. Consider the morphism $rf : a\,F\,h_v \to a\,F\,h_u$. By proposition I - 5, rf is some element in $a\,F\,h_u(v)$. By the construction of the sheaf associated to $F\,h_u$, there is a covering $v = \bigvee_{i \in I} v_i$ in H such that

$rf|_{v_i} \in F\,h_u(v_i)$; in other words, for any $i \in I$ there is a morphism $f_i : F\,h_{v_i} \to F\,h_u$ (proposition I - 4) such that $rf|_{v_i} = af_i$. Now by lemma 9,

$$\text{if } v_i \not\leq u \quad (rf|_{v_i})^{-1}(j_u(S)) = a\,F\,h_{v_i}.$$

Now if $v_i \leq u$, then $rf|_{v_i}$ can be factored into (lemma 8) :

$$a\,F\,h_{v_i} \xrightarrow{\ \alpha\ } a\,F\,h_{v_i} \hookrightarrow a\,F\,h_u.$$

Then, since j is a morphism in $\mathcal{E}(H, \mathbb{T})$, we obtain :

$$\begin{aligned}
&(rf|_{v_i})^{-1}(j_u(S)) \\
&\quad = \alpha^{-1}(j_u(S)|_{v_i}) \\
&\quad = \alpha^{-1}(j_{v_i}(S|_{v_i})) \\
&\quad = j_{v_i}(\alpha^{-1}(S|_{v_i})) \\
&\quad = j_{v_i}(rf|_{v_i})^{-1}(S) \\
&\quad = j_{v_i}(f^{-1}(R \cap S)) \\
&\quad = a\,F\,h_{v_i}.
\end{aligned}$$

Thus we have proved that for any i

$$f^{-1}(R \cap j_u(s))|_{v_i} = (rf|_{v_i})^{-1}(j_u(S)) = a\,F\,h_{v_i}.$$

By lemma 2, this proves that

$$f^{-1}(R \cap j_u(S)) = a\,F\,h_v.$$

Thus any morphism $f : a\,F\,h_v \to R$ factors through $R \cap j_u(S) \rightarrowtail R$. By proposition I - 6, the $a\,F\,h_v$'s determine a proper set of generators; thus $R \cap j_u(S) = R$. Therefore by (L T 2 - 3) :

$$\begin{aligned}
a\,F\,h_u &= j_u(R) \\
&= j_u(R) \cap j_u\,j_u(S) \\
&= a\,F\,h_u \cap j_u(S)
\end{aligned}$$

$= j_u(S)$

which shows that S is in J(u). This implies (G G 3).

Finally consider the assumptions of (G G 4). As j is a morphism in $\mathcal{E}(\mathbb{H}, \mathbb{T})$, we obtain again the following equalities :

$$j_u(R)\big|_{u_i} = j_{u_i}(R\big|_{u_i}) = a\,F\,h_{u_i}.$$

Therefore by lemma 2, $j_u(R) = a\,F\,h_u$ and R is in J(u). This implies (G G 4). ■

Proposition 32.

With the notations of proposition 31, J becomes a subobject of $\Omega_{\mathbb{T}}$ *in* $\mathcal{E}(\mathbb{H}, \mathbb{T})$ *and the following diagram is a pullback in the topos* $\mathcal{E}(\mathbb{H}, \mathbb{T})$

$$\begin{array}{ccc} J & \longrightarrow & 1 \\ \big\downarrow & & \big\downarrow t_{\mathbb{T}} \\ \Omega_{\mathbb{T}} & \xrightarrow[j]{} & \Omega_{\mathbb{T}} \end{array} .$$

If $u \leqslant v$ in $\mathbb{H}$, the restriction mapping $J(v) \to J(u)$ is obtained by pulling back along the canonical inclusion $a\,F\,h_u \to a\,F\,h_v$; this definition makes sense because of axiom (G G 2).

Thus J is already a presheaf on $\mathbb{H}$. This presheaf is separated by lemma 2 applied to $Sh(u{\downarrow}, \mathbb{T}_{u\downarrow})$: if $u = \bigvee_{i\in I} u_i$ and R, S $\in$ J(u) with $R\big|_{u_i} = S\big|_{u_i}$, then $R = \bigcup_{i\in I} R\big|_{u_i}$ and $S = \bigcup_{i\in I} S\big|_{u_i}$; so R = S. Finally (G G 4) says exactly that J is a sheaf.

By definition, J is a subobject of $\Omega_{\mathbb{T}}$ in $Sh(\mathbb{H})$. But by (G G 2) J is stable under action of $M_{\mathbb{T}}$; so J is a subobject of $\Omega_{\mathbb{T}}$ in the topos $Sh(\mathbb{H}, \mathbb{T})$.

Now R $\in$ J(u) if and only if $j_u(R) = a\,F\,h_u = t_{\mathbb{T}}(*)$. This shows that J is the inverse image of $t_{\mathbb{T}}$ along j. ■

Proposition 33.

Consider a Gabriel - Grothendieck $\mathbb{T}$*-topology* J *on the frame* $\mathbb{H}$. *Each* J(u) *is a filter in the lattice of subobjects of* $a\,F\,h_u$.

By (G G 1), a F h_u is in J(u), so J(u) is not empty. Now consider R, S in J(u); we shall prove that R ∩ S is in J(u). For any v in $\mathbb{H}$ and f : a F $h_v \to$ R, consider the following diagram where the squares are pullbacks

$$\begin{array}{ccccc} f^{-1}(R \cap S) & \rightarrowtail & R \cap S & = & S \\ \downarrow & & \downarrow & & \downarrow \\ a\,F\,h_v & \xrightarrow[f]{} & R & \xrightarrow[r]{} & a\,F\,h_u . \end{array}$$

By (G G 2), $f^{-1}(R \cap S) = (rf)^{-1}(s)$ is in J(v) because S is in J(u). Now R ∩ S is some subobject of a F h_u and R is in J(u); moreover for any f : a F $h_v \to$ R, $f^{-1}(R \cap (R \cap S)) = f^{-1}(R \cap S)$ is in J(v); so R ∩ S is in J(u) by (G G 3).

Finally consider R in J(u) and S ⩾ R. For any v in $\mathbb{H}$ and f : a F $h_v \to$ R

$$f^{-1}(R \cap S) = f^{-1}(R) = (rf)^{-1}(R).$$

By (G G 2), $f^{-1}(R \cap S)$ is in J(v) because R is in J(u). Therefore S is in J(u) by (G G 3). ■

§ 8. LOCALIZING AT SOME $\mathbb{T}$-TOPOLOGY

In §§ 5 - 6 - 7, we started from a localization of Sh($\mathbb{H}$, $\mathbb{T}$), where $\mathbb{T}$ is a commutative theory in the topos of sheaves on the frame $\mathbb{H}$, and we constructed successively a universal closure operation on Sh($\mathbb{H}$, $\mathbb{T}$), a Lawvere - Tierney $\mathbb{T}$-topology on $\mathbb{H}$ and a Gabriel - Grothendieck $\mathbb{T}$-topology on $\mathbb{H}$. In this paragraph, we close the loop : from a Gabriel - Grothendieck $\mathbb{T}$-topology on $\mathbb{H}$, we construct a localization of Sh($\mathbb{H}$, $\mathbb{T}$).

The reader may be surprised by the terminology in this paragraph where the words "prelocalized" and "localized" are used instead of the usual terms "separated presheaf" and "sheaf". In fact we are working with two different topologies. The first topology is the canonical one on $\mathbb{H}$: we use the terminology "separated presheaf" and "sheaf" in its usual sense when we refer to the canonical topology on $\mathbb{H}$. But in this paragraph we consider also a $\mathbb{T}$-topology J on $\mathbb{H}$ and thus there will be a corresponding notion of "J-separated-object" in Sh($\mathbb{H}$, $\mathbb{T}$) and "J-sheaf-object" in Sh($\mathbb{H}$, $\mathbb{T}$). To avoid any confusion, we prefer in the latter case to use the words "prelocalized-object" and "localized object".

Thus throughout this paragraph J is a fixed Gabriel - Grothendieck
$\mathbb{T}$-topology on H.

Definition 34.

An object A in Sh(H, $\mathbb{T}$) *is called "prelocalized" (with respect to J)
if for any* u *in* H, R *in* J(u) *and* f, g : a F h_u $\rightrightarrows$ A *we have*

$$f\big|_R = g\big|_R \quad \Rightarrow \quad f = g.$$

Definition 35.

An object A in Sh(H, $\mathbb{T}$) *is called "localized" (with respect to J) if
for any* u *in* H, R *in* J(u) *and* f : R $\rightarrow$ A, *there exists a unique*
g : a F h_u $\rightarrow$ A *such that* g$\big|_R$ = f.

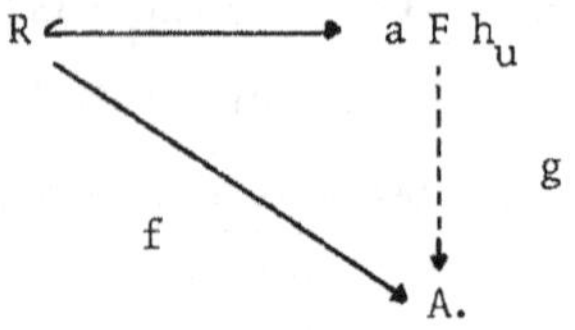

We shall denote by $\mathbb{L}$ the full subcategory of Sh(H, $\mathbb{T}$) whose objects are
the localized ones. It is obvious that any localized object is also prelocalized. We shall prove that $\mathbb{L}$ is a localization of Sh(H, $\mathbb{T}$).

Definition 36.

A monomorphism s : S $\rightarrowtail$ A *in* Sh(H, $\mathbb{T}$) *is called "dense" (with respect
to J) if for any* u *in* H *and any* f : a F h_u $\rightarrow$ A, f^{-1}(S) *is in* J(u).

Proposition 37.

The class of dense monomorphisms is stable under inverse images.

Consider the following diagram where s is dense:

$$
\begin{array}{ccccc}
g^{-1}(f^{-1}(s)) & \longrightarrow & f^{-1}(S) & \longrightarrow & S \\
\big\downarrow{\scriptstyle s''} & {\scriptstyle p.b.} & \big\downarrow{\scriptstyle s'} & {\scriptstyle p.b.} & \big\downarrow{\scriptstyle s} \\
a F h_u & \xrightarrow{\ g\ } & A & \xrightarrow{\ f\ } & B
\end{array}
.
$$

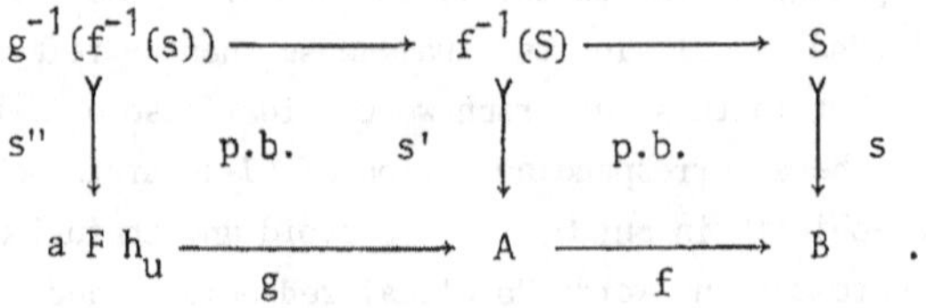

For any $u \in \mathbb{H}$ and $g : a\,F\,h_u \to A$, $g^{-1}(f^{-1}(S)) = (fg)^{-1}(S)$ is in $J(u)$; thus $f^{-1}(S)$ is dense in A. ∎

Proposition 38.

A monomorphism $s : S \rightarrowtail a\,F\,h_u$ *is dense if and only if it is in* $J(u)$.

By (G G 2), any monomorphism in $J(u)$ is a dense monomorphism. Conversely if $S \rightarrowtail a\,F\,h_u$ is dense, choose $R = a\,F\,h_u$ in (G G 3) : this implies exactly that S is in $J(u)$. ∎

Proposition 39.

Consider the composite of two monomorphisms in $Sh(\mathbb{H}, \mathbb{T})$

$$R \overset{r}{\rightarrowtail} S \overset{s}{\rightarrowtail} A.$$

Then $s \circ r$ *is dense if and only if* r *and* s *are dense.*

First suppose r and s are dense. For any u, v in $\mathbb{H}$ and $f : a\,F\,h_u \to A$, $g : a\,F\,h_v \to f^{-1}(S)$, consider the following diagram where all the squares are pullbacks:

$$\begin{array}{ccccccc}
(fg)^{-1}(R) & \longrightarrow & f^{-1}(R) & = & f^{-1}(R) & \longrightarrow & R \\
\downarrow & & \downarrow & & \downarrow & & \downarrow r \\
a\,F\,h_v & \xrightarrow{g} & f^{-1}(S) & = & f^{-1}(S) & \xrightarrow{f'} & S \\
\| & & \| & & \downarrow & & \downarrow s \\
a\,F\,h_v & \xrightarrow{g} & f^{-1}(S) & \rightarrowtail & a\,F\,h_u & \xrightarrow{f} & A .
\end{array}$$

We must prove that $f^{-1}(R) \rightarrowtail a\,F\,h_u$ is in $J(u)$. But $f^{-1}(S) \rightarrowtail a\,F\,h_u$ is in $J(u)$ because s is dense. So by (G G 3) the problem reduces to show that $(fg)^{-1}(R) \rightarrowtail a\,F\,h_v$ is dense. And that is true because r is dense and $(fg)^{-1}(R) \rightarrowtail a\,F\,h_v$ is just $(f'g)^{-1}(r)$.

Conversely suppose $s \circ r$ dense. This implies that for any $f : a\,F\,h_u \to A$, $f^{-1}(R)$ is dense. But $f^{-1}(S) \geqslant f^{-1}(R)$; thus $f^{-1}(S)$ is also dense (proposition 33); therefore s is dense. On the other hand, for any $g : a\,F\,h_v \to S$ consider the following diagram where the squares are pullbacks :

$$\begin{array}{ccccc}
g^{-1}(R) & \longrightarrow & R & = & R \\
\downarrow & & \downarrow r & & \downarrow s \circ r \\
a\,F\,h_v & \xrightarrow{\;g\;} & S & \rightarrowtail_{s} & A
\end{array}$$

$s \circ r$ is dense, thus $g^{-1}(R) = (sg)^{-1}(s \circ r)$ is in $J(v)$; this proves that r is dense. ■

We now turn to the definition of the localizing functor $\ell : Sh(\mathbb{H}, \mathbb{T}) \to \mathbb{L}$. This will be realized in several steps. First we construct a functor $\lambda : Sh(\mathbb{H}, \mathbb{T}) \to Pr(\mathbb{H}, \mathbb{T})$ and we consider the composite $a\lambda a\lambda$ where a is the associated sheaf functor. We prove that this composite functor takes values in $\mathbb{L}$ and we define it to be ℓ.

For any A in $Sh(\mathbb{H}, \mathbb{T})$ and u in $\mathbb{H}$ define

$$\lambda(A)(u) = \varinjlim_{R \in J(u)} (R, A),$$

where the colimit is computed in Sets $^{\mathbb{T}(u)}$. This definition makes sense because (cfr. theorem II - 5)

$$u_! \, u^* R \cong R$$

$$(R, A) \cong (u_! \, u^* R, A) \cong (u^* R, u^* A) \cong (u_! \, u^* R, u_! \, u^* A).$$

Therefore $(R, A) \cong (R|_u, A|_u)$ is provided with the structure of a $\mathbb{T}(u)$-algebra since $\mathbb{T}$ is commutative (lemma 6 applied to $Sh(u{\downarrow}, \mathbb{T}_{u\downarrow})$). Note also that this colimit is filtered (proposition 33).

If $v \leq u$ in $\mathbb{H}$, we need to define a restriction mapping $\lambda(A)(u) \to \lambda(A)(v)$ which is a morphism of $\mathbb{T}(u)$ algebras. But $\lambda(A)(u)$ is defined as a colimit. For any $R \in J(u)$, $R|_v$ is in $J(v)$ by (G G 2) and by composition with $R|_v \hookrightarrow R$ we obtain a $\mathbb{T}(u)$-homomorphism

$$(R, A) \to (R|_v, A) \xrightarrow{\;s^v_{R|_v}\;} \lambda(A)(v),$$

where the second morphism is the canonical inclusion into the colimit $\lambda(A)(v)$. If $S \leq R$ in $J(u)$, the following diagram is clearly commutative:

$$\begin{array}{ccccc} (R,\ A) & \longrightarrow & (R|_v,\ A) & \xrightarrow{s^v_{R|_v}} & \\ \downarrow & & \downarrow & & \lambda(A)(v) \\ (S,\ A) & \longrightarrow & (S|_v,\ A) & \xrightarrow[s^v_{S|_v}]{} & \end{array}$$

where the vertical arrows are the composition with the canonical inclusions $S \rightarrowtail R$ and $S|_v \rightarrowtail R|_v$. Therefore we have a cone and thus a unique factorization $\lambda(A)(u) \to \lambda(A)(v)$. It is obvious that this makes $\lambda(A)$ into a $\mathbb{T}$-presheaf.

Finally, consider a morphism $f : A \to B$ in $\mathrm{Sh}(\mathbf{H}, \mathbb{T})$. We must define $\lambda f : \lambda A \to \lambda B$, thus for any u in $\mathbf{H}$, a $\mathbb{T}(u)$-morphism $\lambda f(u) : \lambda A(u) \to \lambda B(u)$. But again $\lambda A(u)$ is defined by a colimit. Consider R in $J(u)$ and the following morphism

$$(R,\ A) \xrightarrow{(1,\ f)} (R,\ B) \xrightarrow{s^u_R} \lambda(B)(u).$$

If $S \leqslant R$ in $J(u)$ and $s : S \rightarrowtail R$ is the inclusion, the following diagram is commutative

$$\begin{array}{ccccc} (R,\ A) & \xrightarrow{(1,\ f)} & (R,\ B) & \xrightarrow{s^u_R} & \\ {\scriptstyle (s,\ 1)}\downarrow & & \downarrow{\scriptstyle (s,\ 1)} & & \lambda(B)(u) \\ (S,\ A) & \xrightarrow[(1,\ f)]{} & (S,\ B) & \xrightarrow[s^u_S]{} & \end{array} .$$

Therefore there exist a unique factorization $\lambda(f)(u) : \lambda(A)(u) \to \lambda(B)(u)$ through the colimit $\lambda(A)(u)$. Clearly this makes λ into a functor $\lambda : \mathrm{Sh}(\mathbf{H}, \mathbb{T}) \to \mathrm{Pr}(\mathbf{H}, \mathbb{T})$.

Proposition 40.

For any A *in* $\mathrm{Sh}(\mathbf{H}, \mathbb{T})$, λA *is a separated presheaf.*

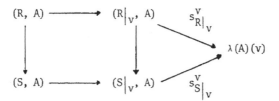

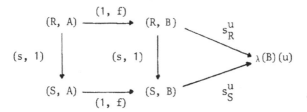

Consider $u = \bigvee_{i \in I} u_i$ in $\mathbf{H}$ and x, y in $\lambda(A)(u)$ such that, for any i in I, $x|_{u_i} = y|_{u_i}$. We must prove that $x = y$. $\lambda(A)(u)$ is a filtered colimit; therefore there exists some R in $J(u)$ and $\overline{x}, \overline{y} : R \rightrightarrows A$ such that x and y are represented by $\overline{x}$, $\overline{y}$ in the colimit. From $x|_{u_i} = y|_{u_i}$, we deduce that $\overline{x}|_{u_i}$ and $\overline{y}|_{u_i}$ represent the same element in the colimit $\lambda A(u_i)$; this means that there exists some $R_i \in J(u_i)$ such that $(\overline{x}|_{u_i})|_{R_i} = (\overline{y}|_{u_i})|_{R_i}$ or, looking at R_i as a subobject of R, $\overline{x}|_{R_i} = \overline{y}|_{R_i}$. Denote by K the equalizer of $\overline{x}$, $\overline{y}$; from $\overline{x}|_{R_i} = \overline{y}|_{R_i}$ we deduce that $R_i \leqslant K$.

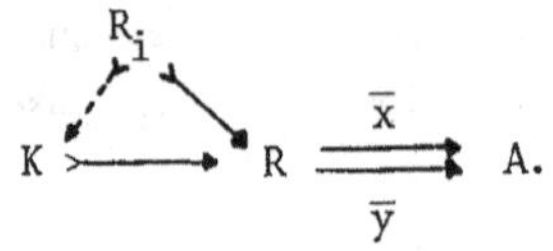

Thus for any $i \in I$, $K|_{u_i} \geqslant R_i|_{u_i} = R_i$ and from $R_i \in J(u_i)$ we deduce that $K|_{u_i} \in J(u_i)$ (proposition 33). By (G G 4) this implies that $K \in J(u)$. But $\overline{x}$ and $\overline{y}$ coincide on K, so $x = y$ in $\lambda A(u)$ by construction of filtered colimits in algebraic categories, (cfr. [21] - 18 - 3 - 6). ■

For any object A in Sh(H, T), we denote by $\lambda_A : A \to \lambda A$ the morphism whose component at $u \in H$ is given by (cfr. proposition I - 5)

$$A(u) \cong (a\, F\, h_u, A) \xrightarrow[s^u_{a F h_u}]{} \lambda A(u).$$

If P is some presheaf, we also denote by $a_P : P \to aP$ the canonical morphism arising from the construction of the associated sheaf functor.

<u>Proposition 41.</u>

For any $u \in H$, $R \in J(u)$, $A \in$ Sh(H, T), $\overline{x} : R \to A$ *and* $x : a\, F\, h_u \to a\lambda A$, *the following square is commutative*

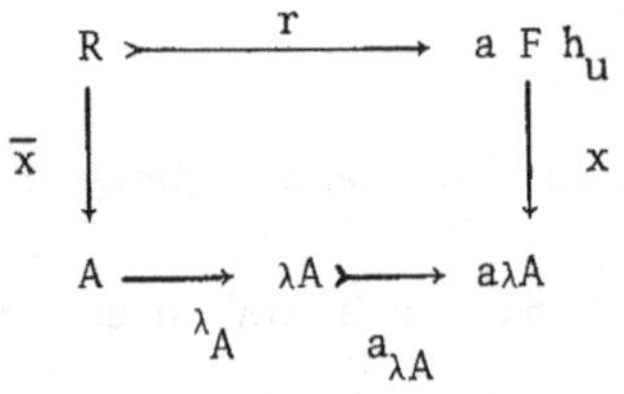

if and only if x *determines (via proposition* I - 5*) an element in* $\lambda A(u)$ *represented by* $\overline{x} \in (R, A)$ *in the filtered colimit.*

By proposition 40, λA is separated and thus $a_{\lambda A}$ is a monomorphism. Now for any $v \in H$ and $f : a\,F\,h_v \to \lambda A$ in $Pr(H, \mathbb{T})$, by propositions I 4 - I 5 $f \circ a_{F\,h_v} : F\,h_v \to a\,F\,h_v \to \lambda A$ determines an element in $\lambda A(v) \subseteq a\lambda A(v)$ and $a_{\lambda A} \circ f : a\,F\,h_v \to \lambda A \rightarrowtail a\lambda A$ determines an element in $a\lambda A(v)$; this is clearly the same element as the one determined in $Sh(H, \mathbb{T})$ by

$$a_{\lambda A} \circ f \circ a_{F\,h_v} : F\,h_v \to a\,F\,h_v \to \lambda A \rightarrowtail a\lambda A.$$

Suppose first that $x = ax'$ where $x' : F\,h_u \to \lambda A$ determines some element in $\lambda A(u)$ and moreover that this element is represented by $\overline{x}$ in the colimit $\lambda A(u)$. By theorem II - 5, $R \cong u_!\, u^* R = R|_u$ and thus any morphism $R \to B$ in $Sh(H, \mathbb{T})$ factors through the canonical monomorphism $B|_u \rightarrowtail B$. Thus the commutativity of the diagram is equivalent to the commutativity of its restriction at u.

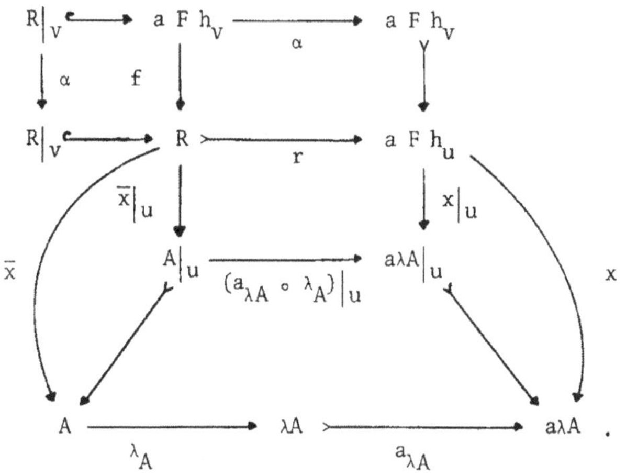

But the diagram restricted at u is commutative if and only if it is commutative when preceded by any morphism $f : a\,F\,h_v \to R$, with $v \leqslant u$ (proposition I - 6 applied to $Sh(u{\downarrow}, \mathbb{T}_{u\downarrow})$. Finally, composing again with the monomorphism $a\lambda A|_u \rightarrowtail a\lambda A$, the diagram we need to consider is commutative if and only if it is commutative when preceded by any $f : a\,F\,h_v \to R$, with $v \leqslant u$.

So, consider $v \leqslant u$ in H and $f : a\,F\,h_v \to R$. Via proposition I - 5, $a_{\lambda A} \circ \lambda_A \circ \overline{x} \circ f \circ a_{F\,h_v}$ determines some element in $\lambda A(v)$. By definition of λ_A, this is the element represented by $\overline{x}f \in (a\,F\,h_v, A)$. Thus $a_{\lambda A} \circ \lambda_A \circ \overline{x} \circ f$

determines in fact an element in $\lambda A(u)$ represented by $\bar{x}f \in (a\ F\ h_v,\ A)$. On the other hand x is represented in $\lambda A(u)$ by $\bar{x} \in (R,\ A)$. We must consider $x \circ r \circ f$. But rf can be factored by some $\alpha : a\ F\ h_v \to a\ F\ h_v$ through the canonical inclusion : $a\ F\ h_v \hookrightarrow a\ F\ h_u$ (lemma 8). Sincc λA is a presheaf, the element of $\lambda A(v)$ determined by $x|_v$ is represented by $\bar{x}|_v \in (R|_v,\ A)$. Now the construction of λA is compatible with the algebraic $\mathbb{T}$-structure and in particular with the action of the 1-ary operation α (cfr. lemma 7). Therefore the element of $\lambda A(v)$ determined by $x|_v \circ \alpha$ is represented by $\bar{x}|_v \circ \alpha \in (R|_v,\ A)$ where we still denote by $\alpha : R|_v \to R|_v$, the action of the 1-ary operation α on $R|_v$. So we must prove that $\bar{x}\ f \in (a\ F\ h_v,\ A)$ and $\bar{x}|_v \circ \alpha \in (R|_v,\ A)$ represent the same element in the colimit $\lambda A(v)$.

Consider the following commutative diagram

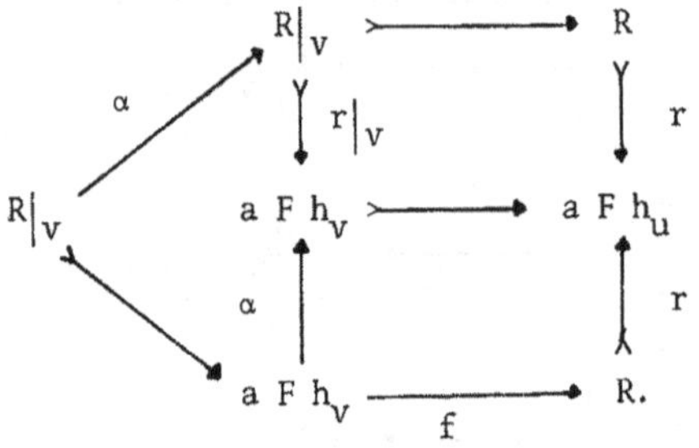

Since r is a monomorphism, we obtain the commutativity of the following diagram

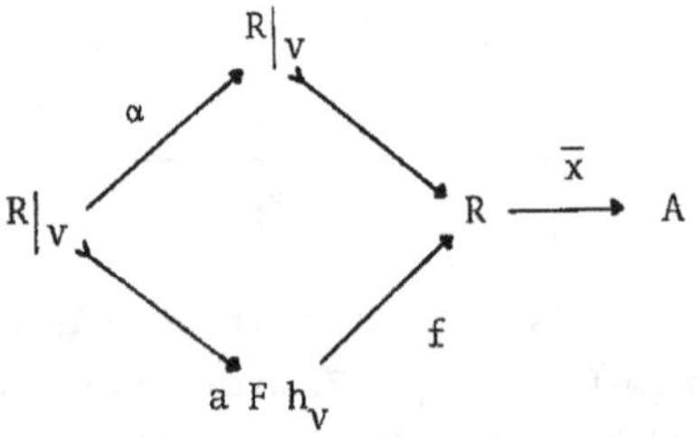

which shows that $\bar{x}|_v \circ \alpha$ and $\bar{x}\ f$ coincide on $R|_v$ which is in $J(v)$. So $\bar{x}|_v \circ \alpha$ and $\bar{x}\ f$ represent the same element in $\lambda A(v)$. This concludes the proof of the commutativity of the given diagram.

Conversely suppose the diagram to be commutative. $\bar{x} \in (R,\ A)$ represents some element in $\lambda A(u)$ determined by a morphism $x' : F\ h_u \to \lambda A$ in $Pr(\mathbb{H}, \mathbb{T})$ (proposition I - 4). So, by the first part of the proof, the following diagram is commutative

$$\begin{array}{ccccc} R & & \xrightarrow{\quad r \quad} & & a\,F\,h_u \\ {\scriptstyle \overline{x}}\downarrow & & & & \downarrow{\scriptstyle ax'} \\ A & \xrightarrow[\lambda_A]{} & \lambda A & \xrightarrow[a_{\lambda A}]{} & a\lambda A\,. \end{array}$$

But on the other hand λA is separated (proposition 40) and thus by construction of the associated sheaf functor, there is a covering $u = \bigvee_{i \in I} u_i$ in H such that for any i, $x|_{u_i}$ determines an element in $\lambda A(u_i)$; thus the following diagram is commutative

$$\begin{array}{ccccc} R|_{u_i} & & \xrightarrow{\quad r_i \quad} & & a\,F\,h_{u_i} \\ {\scriptstyle \overline{x}|_{u_i}}\downarrow & & & & \downarrow{\scriptstyle x|_{u_i}} \\ A & \xrightarrow[\lambda_A]{} & \lambda A & \xrightarrow[a_{\lambda A}]{} & a\lambda A \end{array}$$

where $x|_{u_i}$ represents some element in $\lambda A(u_i)$. ax' is represented by $\overline{x}$ and thus $ax'|_{u_i}$ is represented by $\overline{x}|_{u_i}$. If we can prove that $x|_{u_i}$ is also represented by $\overline{x}|_{u_i}$, then for any i, $x|_{u_i} = ax'|_{u_i}$ and by lemma 2, $x = ax'$. Then we can reduce the problem to the case where x represents some element in $\lambda A(u)$. For such an element x, we must prove that $\overline{x}$ represents x in $\lambda A(u)$. Consider $\tilde{x} \in (\tilde{R}, A)$ which represents x in $\lambda A(u)$; there is no loss of generality to suppose $R = \tilde{R}$, (if not, simply work on $R \cap \tilde{R}$ which is still in J(u) by proposition 33). So, by the first part of the proof, we have two commutative diagrams

$$\begin{array}{ccccc} R & & \xrightarrow{\quad r \quad} & & a\,F\,h_u \\ {\scriptstyle \overline{x}}\downarrow & & & & \downarrow{\scriptstyle x} \\ A & \xrightarrow[\lambda_A]{} & \lambda A & \xrightarrow[a_{\lambda A}]{} & a\lambda A \end{array} \qquad \begin{array}{ccccc} R & & \xrightarrow{\quad r \quad} & & a\,F\,h_u \\ {\scriptstyle \tilde{x}}\downarrow & & & & \downarrow{\scriptstyle x} \\ A & \xrightarrow[\lambda_A]{} & A & \xrightarrow[a_{\lambda A}]{} & a\lambda A \end{array}$$

and $\tilde{x}$ represents x in $\lambda A(u)$. It suffices to prove that $\overline{x}$ and $\tilde{x}$ represent the same element in $\lambda A(u)$, i.e. that $\bar{x}$ and $\tilde{x}$ coincide on some $K \leqslant R$ with $K \in J(u)$. Take K to be the equalizer of $\bar{x}$ and $\tilde{x}$. $\bar{x}$ and $\tilde{x}$ coincide on K and it suffices to prove that $K \in J(u)$. We have the following situation

$$K \overset{k}{\rightarrowtail} R \overset{r}{\rightarrowtail} a\,F\,h_u$$

with r dense. By proposition 39, $r\,k$ will be dense and thus K will be in $J(u)$ by proposition 38, as soon as k is a dense monomorphism. To verify the latter, consider $v \in H$ and $f : a\,F\,h_v \rightarrow R$. We must prove that $f^{-1}(K)$ is in $J(v)$. In the following diagram

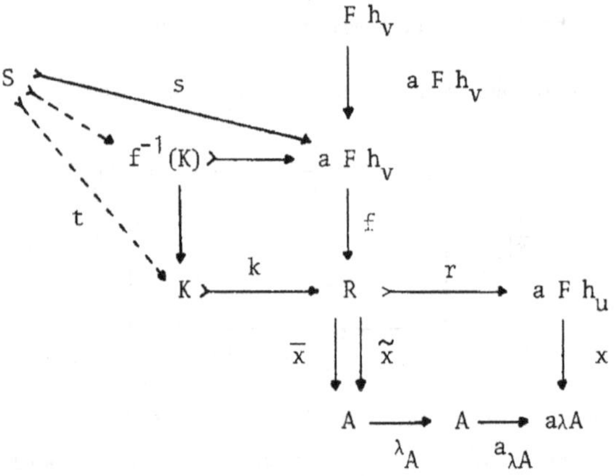

we have :

$a_{\lambda A} \circ \lambda_A \circ \bar{x} \circ f = x \circ r \circ f = a_{\lambda A} \circ \lambda_A \circ \tilde{x} \circ f$. But λA is a separated presheaf (proposition 39); thus $a_{\lambda A}$ is a monomorphism and $\lambda_A \circ \bar{x} \circ f = \lambda_A \circ \tilde{x} \circ f$.

Thus $\lambda_A \circ \bar{x} \circ f \circ a_{F\,h_v}$ and $\lambda_A \circ \tilde{x} \circ f \circ a_{F\,h_v}$ determine the same element of $\lambda A(v)$. By definition of λ_A, this element is thus represented by

$$\bar{x} \circ f \in (a\,F\,h_v, A) \text{ and } \tilde{x} \circ f \in (a\,F\,h_v, A).$$

By construction of the filtered colimit, this means that there exists some $s : S \rightarrowtail a\,F\,h_v$ in $J(v)$ such that $\bar{x} \circ f \circ s = \tilde{x} \circ f \circ s$. But then $f \circ s$ factors through the equalizer k of $(\bar{x}, \tilde{x})$. Let t be the factorization morphism. Finally s and t factor through the pullback $f^{-1}(K)$. This shows that $S \leqslant f^{-1}(K)$. But S is in $J(v)$ and thus by proposition 33, $f^{-1}(K)$ is in $J(v)$. ∎

<u>Proposition 42.</u>

For any A *in* $\mathrm{Sh}(H, \mathbb{T})$, $a\lambda(A)$ *is a prelocalized object.*

Consider $u \in \mathbb{H}$, $r : R \rightarrowtail a\,F\,h_u$ in $J(u)$ and $f, g : a\,F\,h_u \rightrightarrows a\lambda(A)$ such that $fr = gr$. We must prove that $f = g$. Now by proposition I - 5, f is an element in $a\lambda A(u)$ and since λA is a separated presheaf (proposition 40), the construction of the associated sheaf functor shows that there is a covering $u = \bigvee_{i \in I} u_i$ in $\mathbb{H}$ such that for any $i \in I$, $f|_{u_i} \in \lambda A(u_i)$. Thus $f|_{u_i}$ can be represented by some $\overline{f}_i \in (R_i, A)$ with R_i in $J(u_i)$. A similar thing can be done about g and again without any loss of generality we may assume that the covering working for g is the same as the one for f (eventually consider a common refinement of both coverings). We may also assume that the corresponding subobjects R_i agree (eventually consider the intersection of both subobjects). Thus for any $i \in I$, $g|_{u_i} \in \lambda A(u_i)$ is represented by $\overline{g}_i \in (R_i, A)$.

Consider the following commutative diagram (by proposition 41) where $a_{\lambda A}$ is the canonical morphism arising from the construction of the associated sheaf functor. $a_{\lambda A}$ is a monomorphism because λA is separated (proposition 40).

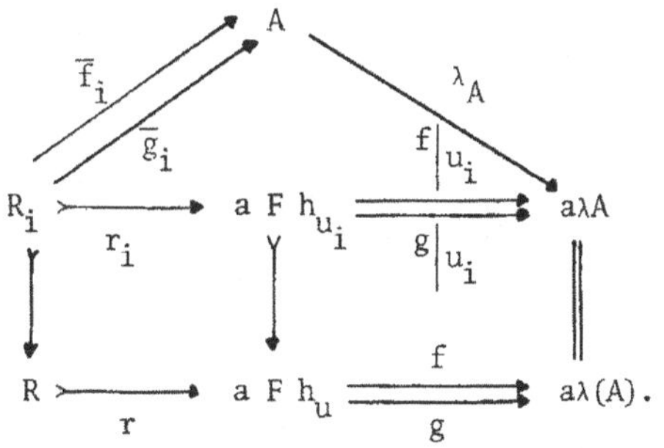

From $fr = gr$, we deduce that

$$\lambda_A \circ \overline{f}_i = \lambda_A \circ \overline{g}_i .$$

Thus the following diagrams are commutative

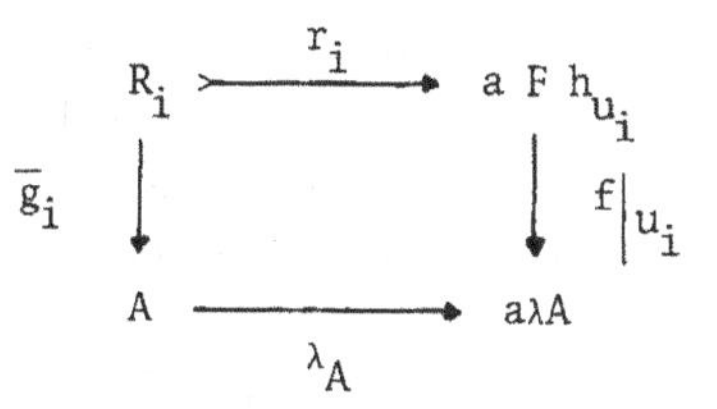

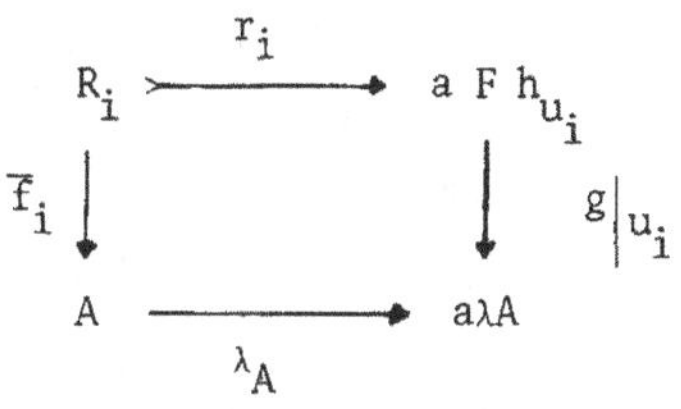

and thus, by proposition 41, $\overline{g}_i$ represents $f|_{u_i}$ in $\lambda A(u_i)$ and $\overline{f}_i$ represents $g|_{u_i}$

in $\lambda A(u_i)$. But $\overline{f}_i$ represents $f|_{u_i}$ and $\overline{g}_i$ represents $g|_{u_i}$. So $f|_{u_i} = g|_{u_i}$ for any i and thus f = g, since λA is a separated presheaf (proposition 40). ∎

Proposition 43.

For any prelocalized object A in Sh(H, T), the composite $a\lambda_A \circ \lambda_A : A \to \lambda A \to a\lambda A$ *is a dense monomorphism.*

$a_{\lambda A}$ is a monomorphism because λA is a separated presheaf (proposition 40). To verify that λ_A is a monomorphism, consider $f, g : B \rightrightarrows A$ such that $\lambda_A \circ f = \lambda_A \circ g$. For any $u \in H$ and

$k : a\,F\,h_u \to B$, $a_{\lambda A} \circ \lambda_A \circ (f\,k) = a_{\lambda A} \circ \lambda_A \circ (g\,k)$; by proposition 41, this implies (consider the following diagram where $a\,F\,h_u \in J(u)$)

$$\begin{array}{ccccc} a\,F\,h_u & & = & & a\,F\,h_u \\ fk \downarrow\downarrow gk & & & & \downarrow \; a_{\lambda A} \circ \lambda_A \circ f \circ k = a_{\lambda A} \circ \lambda_A \circ g \circ k \\ A & \xrightarrow{\lambda_A} & \lambda A & \xrightarrow{a_{\lambda A}} & a\lambda A \end{array}$$

that fk and gk represent the same element $\lambda_A\,fk = \lambda_A\,gk$ of $\lambda A(u)$. By construction of a filtered colimit, this means that fk and gk coincide on some $R \in J(u)$.

$$R \overset{r}{\rightarrowtail} a\,F\,h_u \overset{fk}{\underset{gk}{\rightrightarrows}} A.$$

But A is prelocalized, thus from fkr = gkr, we deduce fk = gk. By proposition I - 6, we conclude that f = g. Thus λ_A is a monomorphism.

Now we will establish that $a_{\lambda A} \circ \lambda_A$ is a dense monomorphism in Sh(H, T). (Note that the monomorphisms $a_{\lambda A}$ and λ_A are not in Sh(H, T) but only in Pr(H, T)). For any $u \in H$ and $f : a\,F\,h_u \to a\lambda A$, we must prove that $f^{-1}(a_{\lambda A} \circ \lambda_A)$ is in J(u). But f determines an element in $a\lambda A(u)$ (proposition I - 5) and by construction of the associated sheaf functor (via proposition 40), there is a covering $u = \bigvee_{i \in I} u_i$ in H such that each $f|_{u_i}$ determines some element in $\lambda A(u_i)$. This implies that $f|_{u_i}$ is represented by some $\overline{f}_i \in (R_i, A)$ with $R_i \in J(u_i)$ and making the following diagram

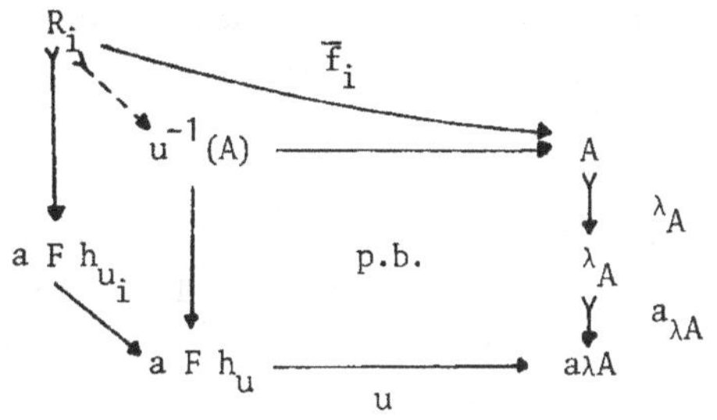

where the square is a pullback, commute (proposition 41). This implies that $R_i \leqslant u^{-1}(A)$ and since $R_i \in J(u_i)$ we have $u^{-1}(A) \in J(u_i)$, (proposition 33). ■

Proposition 44.

If A *is a prelocalized object in* Sh(H, T), $a\lambda A$ *is a localized object.*

Consider $u \in H$, $R \in J(u)$ and $f : R \to a\lambda A$. We need to find an extension $\tilde{f} : a\,F\,h_u \to a\lambda A$ of f. This extension will be necessarily unique because $a\lambda A$ is prelocalized (proposition 42).

Consider the following diagram where the squares are pullbacks and $\tilde{f}$ as defined below

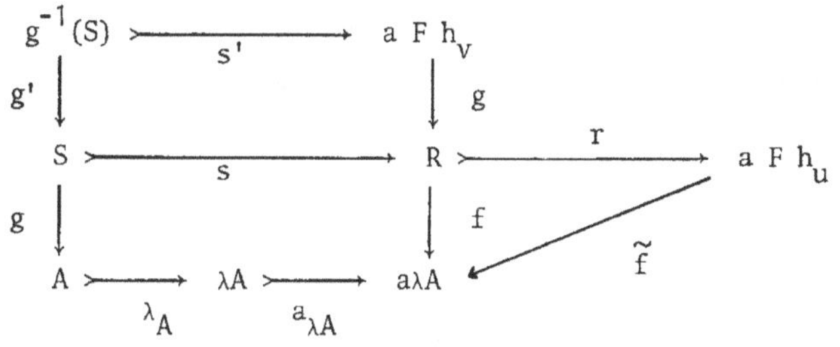

Now $a_{\lambda A} \circ \lambda_A$ is dense, thus s is dense (proposition 37), thus $r \circ s$ is dense (propositions 38 - 39). Therefore $g \in (S, A)$ represents some element in $\lambda A(u)$; this element is determined (proposition 41) by a morphism $\tilde{f} : a\,F\,h_u \to a\lambda A$ such that $\tilde{f} \circ r \circ s = a_{\lambda A} \circ \lambda_A \circ g$.

We must prove that $\tilde{f} \circ r = f$. By proposition I - 6, it suffices to show that for any $v \in H$ and $g : a\,F\,h_v \to R$, $\tilde{f} \circ r \circ g = f \circ g$. Forming the pullback of g and s, we obtain

$$\tilde{f} \circ r \circ g \circ s' = f \circ g \circ s'$$

because $\tilde{f} \circ r$ and f coincide on s. By proposition 37, s' is dense and thus in J(v) by proposition 38. But aλA is prelocalized; so $\tilde{f} \circ r \circ g = f \circ g$. ∎

Definition 45.

If J *is a* 𝕋*-topology on* ℍ *and* 𝕃 *is the full subcategory of localized objects of* Sh(ℍ, 𝕋)*, the functor* ℓ : Sh(ℍ, 𝕋) → 𝕃 *is defined by* ℓ = aλaλ.

By propositions 42 - 44, each aλaλA is localized and thus this definition makes sense. We shall prove that ℓ is a localization.

Proposition 46.

Let f : A → B *be a morphism in* Sh(ℍ, 𝕋) *with* B *a localized object. Then* f *factors uniquely through* λ_A.

We must define g in Pr(ℍ, 𝕋) such that the following diagram commutes

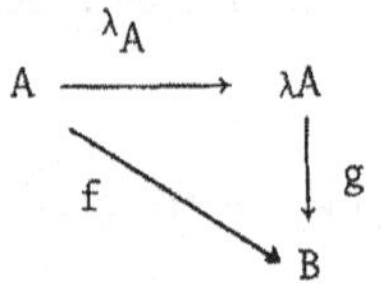

For any u ∈ ℍ, we must define g(u) : λA(u) → B(u). Consider some element x ∈ λA(u) represented by $\bar{x}$ ∈ (R, A) with R ∈ J(u). From the following situation

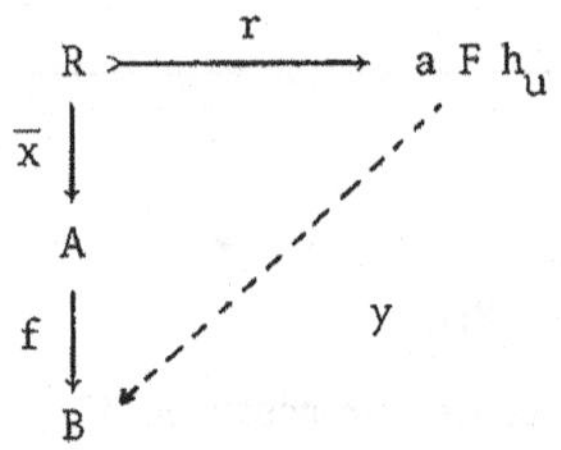

and the fact that B is localized, we get a unique extension y such that $y \circ r = f \circ \bar{x}$. Thus y determines some element in B(u) : this is g(u)(x).

This definition does not depend on the choice of $\bar{x}$ representing x. If $\bar{x}'$ ∈ (R', A) is another element representing x, $\bar{x}$ and $\bar{x}'$ coincide on some R'' ∈ J(u) and thus the corresponding extensions y, y' coincide also on R'' ∈ J(u); because B is localized, this implies that y = y'.

Now if $\alpha \in \mathcal{O}_n(u)$ is some operation and if $x_1, \ldots, x_n \in \lambda A(u)$ are represented by the morphisms $\overline{x}_1, \ldots, \overline{x}_n$, without any loss of generality, we may suppose that $\overline{x}_1, \ldots, \overline{x}_n$ are defined on the same $R \in J(u)$ - (if not, take their intersection). Denote by y_i the unique extension of $f \circ \overline{x}_i$ to $a\,F\,h_u$. From the commutativity of

$$\begin{array}{ccc} R & \overset{r}{\rightarrowtail} & a\,F\,h_u \\ {\scriptstyle \overline{x}_i}\downarrow & & \downarrow{\scriptstyle y_i} \\ A & \underset{f}{\longrightarrow} & B \end{array}$$

and the fact that f is a $\mathbb{T}$-homomorphism, we deduce the commutativity of

$$\begin{array}{ccc} R & \rightarrowtail & a\,F\,h_u \\ {\scriptstyle \alpha(\overline{x}_1, \ldots, \overline{x}_n)}\downarrow & & \downarrow{\scriptstyle \alpha(y_1, \ldots, y_n)} \\ A & \underset{f}{\longrightarrow} & B \end{array}$$

(this makes sense because R and $a\,F\,h_u$ are in $Sh(u{\downarrow}, \mathbb{T}_{u\downarrow})$). This shows that $g(u)$ is a $\mathbb{T}(u)$-homomorphism.

Finally if $v \leqslant u$ in $\mathbb{H}$ and $x \in \lambda A(u)$ is represented by $\overline{x} \in (R, A)$ then $x|_v \in \lambda A(v)$ is represented by $\overline{x}|_v$. Consider the following diagram

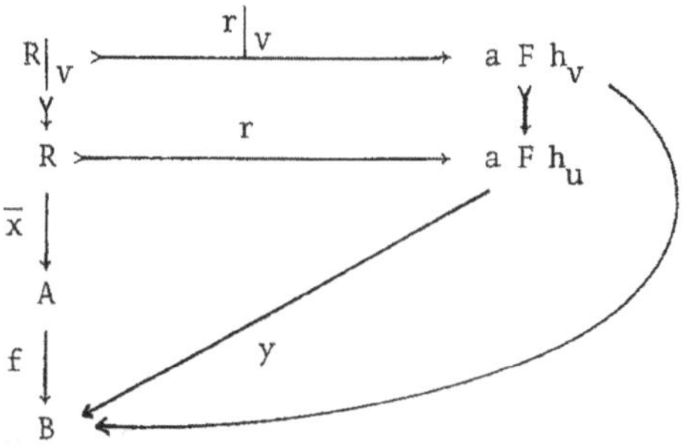

where $y = g(u)(x)$ and $z = g(v)(x|_v)$. This diagram shows that $y|_v$ and z coincide on $R|_v$; because B is localized, $z = y|_v$ or in other words, $g(u)(x)|_v = g(v)(x|_v)$. This proves that $g : \lambda A \to B$ is a morphism in $Pr(\mathbb{H}, \mathbb{T})$.

Now for an element $x \in A(u)$, $\lambda_A(u)(x)$ is represented by $x \in (a\,F\,h_u, A)$ and thus the unique extension y is necessarily $f \circ x$

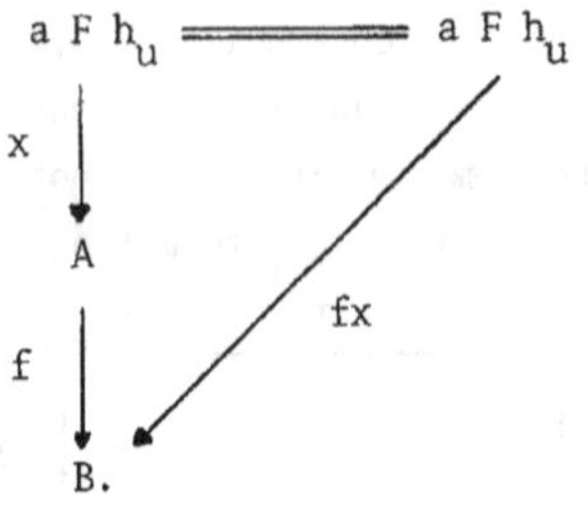

This shows that $g \circ \lambda_A = f$ and concludes the proof. ■

Theorem 47.

Let J *be a Gabriel - Grothendieck* $\mathbb{T}$*-topology on* $\mathbb{H}$. *The situation*
$\mathbb{L} \underset{i}{\overset{\ell}{\leftrightarrows}} \mathrm{Sh}(\mathbb{H}, \mathbb{T})$ *is a localization of* $\mathrm{Sh}(\mathbb{H}, \mathbb{T})$.

We must prove that ℓ is left exact and left adjoint to the canonical inclusion i. But $\ell = a\lambda a\lambda$ and a is left exact. Moreover, λ is also left exact because finite limits are computed pointwise in $\mathrm{Sh}(\mathbb{H}, \mathbb{T})$ and $\mathrm{Pr}(\mathbb{H}, \mathbb{T})$ and $\lambda A(u)$ is defined by a filtered colimit in $\underline{\mathrm{Sets}}^{\mathbb{T}(u)}$ where filtered colimits commute with finite limits (cfr. [21] - 18 - 3 - 6). Finally, ℓ is left exact.

Now consider the following situation with A, B in $\mathrm{Sh}(\mathbb{H}, \mathbb{T})$ and B localized.

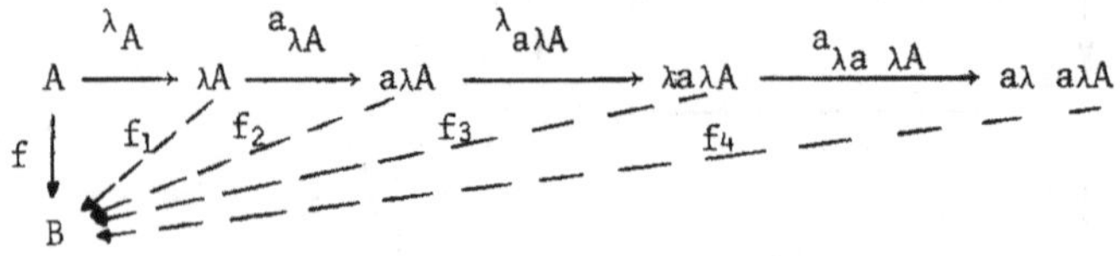

There is a unique extension f_1 by proposition 46, a unique extension f_2 because B is a sheaf, a unique extension f_3 by proposition 46 and finally a unique extension f_4 because B is a sheaf. This shows that the morphism $A \to a\lambda a\lambda A$ has the universal property making ℓ left adjoint to i. ■

§ 9. CLASSIFICATION OF THE LOCALIZATIONS OF $\mathrm{Sh}(\mathbb{H}, \mathbb{T})$

In § 5 - 6 - 7 - 8, we have described correspondances between localizations of $\mathrm{Sh}(\mathbb{H}, \mathbb{T})$, universal closure operations on $\mathrm{Sh}(\mathbb{H}, \mathbb{T})$, Lawvere - Tierney $\mathbb{T}$-topologies on $\mathbb{H}$ and Gabriel - Grothendieck $\mathbb{T}$-topologies on $\mathbb{H}$. In this paragraph, we show that all these correspondances are bijective and thus we get a three-fold classification of the localizations of Sh($\mathbb{H}$, $\mathbb{T}$).

Proposition 48.

Let $\mathbb{L} \underset{i}{\overset{\ell}{\leftrightarrows}}$ Sh($\mathbb{H}$, $\mathbb{T}$) *be a localization of* Sh($\mathbb{H}$, $\mathbb{T}$) *and* $f : A \to B$ *a morphism in* Sh($\mathbb{H}$, $\mathbb{T}$).
Then f *is carried by* ℓ *to an isomorphism*

if and only if

the image of f *and the equalizer of the kernel pair of* f *are carried by* ℓ *to isomorphisms.*

This follows easily from the fact that ℓ is a right and left exact functor between regular categories. ■

Proposition 49.

Let $\mathbb{L} \underset{i}{\overset{\ell}{\leftrightarrows}}$ Sh($\mathbb{H}$, $\mathbb{T}$) *be a localization of* Sh($\mathbb{H}$, $\mathbb{T}$) *and* $r : R \rightarrowtail A$ *a monomorphism in* Sh($\mathbb{H}$, $\mathbb{T}$).
Then r *is applied by* ℓ *on an isomorphism*

if and only if

for any u *in* $\mathbb{H}$ *and* $f : a\,F\,h_u \to A$, $f^{-1}(r)$ *is sended by* ℓ *to an isomorphism.*

The downward implication is obvious since ℓ commutes with inverse images. Now suppose that for any u in $\mathbb{H}$ and $f : a\,F\,h_u \to A$, $\ell f^{-1}(r)$ is an isomorphism. Consider the following composite

$$\coprod_{\substack{v \in \mathbb{H} \\ f : a F h_v \to A}} f^{-1}(R) \xrightarrow{\ q\ } \coprod_{\substack{v \in \mathbb{H} \\ f : a F h_v \to A}} a\,F\,h_v \xrightarrow{\ p\ } A$$

where q acts by inverse image on each component and p is the canonical regular epimorphism whose existence is implied by proposition I - 6. Each monomorphism

$$f^{-1}(R) \rightarrowtail a\,F\,h_v$$

is sended by ℓ to an isomorphism. Thus $\ell(q)$ is an isomorphism. But $p \circ q$ obviously factors through R via the morphism $f^{-1}(R) \to R$. So we have the following commutative diagram in $\mathbb{L}$

$$\begin{array}{ccccc} & & \ell R & & \\ & \nearrow & & \searrow^{\ell r} & \\ \cdot & & & & \ell A \\ & {}_{\ell q}\searrow^{\cong} & & \nearrow_{\ell p} & \\ & & \cdot & & \end{array}$$

Now $\ell p \circ \ell q$ is a regular epimorphism; so ℓr is both a monomorphism and a regular epimorphism; thus it is an isomorphism. ■

Proposition 50.

Consider a localization $\mathbb{L} \underset{i}{\overset{\ell}{\rightleftarrows}} \mathrm{Sh}(\mathbb{H}, \mathbb{T})$ *and the corresponding closure operation. A monomorphism* $r : R \rightarrowtail A$ *in* $\mathrm{Sh}(\mathbb{H}, \mathbb{T})$ *is sended by* ℓ *to an isomorphism if and only if* $\overline{R} = A$.

$\overline{R}$ is defined by the following pullback, where $A \to \ell A$ is the canonical morphism arising from the adjunction

$$\begin{array}{ccc} \overline{R} & \longrightarrow & \ell R \\ \big\downarrow & \text{p.b.} & \big\downarrow \ell r \\ A & \longrightarrow & \ell A. \end{array}$$

Obviously if ℓr is an isomorphism, $\overline{R} = A$. Conversely, if $\overline{R} = A$ apply ℓ to this diagram. Since ℓ is idempotent, we get

$$\begin{array}{ccc} \ell A & \longrightarrow & \ell R \\ \big\| & & \big\downarrow \ell r \\ \ell A & = & \ell A. \end{array}$$

This shows that ℓr is a regular epimorphism and thus an isomorphism. ■

Proposition 51.

Consider a Gabriel - Grothendieck $\mathbb{T}$*-topology* J *on* $\mathbb{H}$ *and the corresponding localization* $\mathbb{L} \underset{i}{\overset{\ell}{\rightleftarrows}} \mathrm{Sh}(\mathbb{H}, \mathbb{T})$ *of* $\mathrm{Sh}(\mathbb{H}, \mathbb{T})$. *A monomorphism* $r : R \rightarrowtail a\, F\, h_u$ *is sended by* ℓ *to an isomorphism if and only if it is in* $J(u)$.

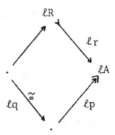

Suppose R in J(u). For any localized object A, we have the natural isomorphisms

$$(\ell R, A) \cong (R, A) \cong (a\ F\ h_u, A) \cong (\ell\ a\ F\ h_u, A)$$

which prove that $\ell R \cong \ell\ a\ F\ h_u$.

Conversely suppose ℓr to be an isomorphism. Consider the following diagram where the upper square is a pullback

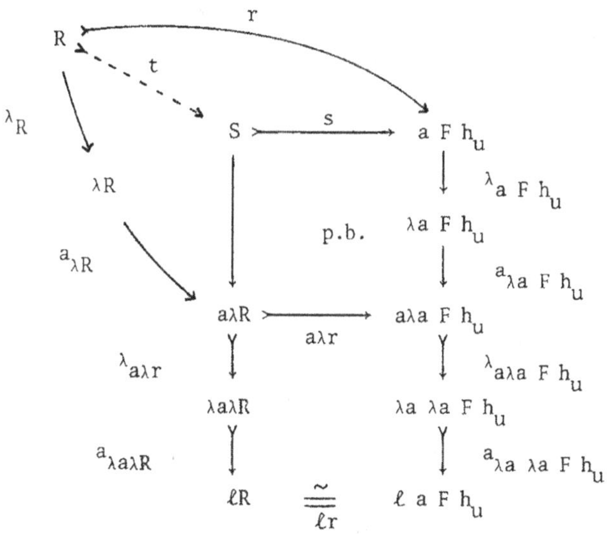

By propositions 37, 38, 39, 42, 43, $a_{\lambda a \lambda R} \circ \lambda_{a\lambda}$ is dense and thus $a\lambda r$ is dense; therefore s is dense. Now the commutativity of the diagram gives rise to a monomorphic factorization $t : R \rightarrowtail S$ and to prove that $r = s \circ t$ is dense, it suffices to prove that t is dense.

So consider $v \in H$ and $g : a\ F\ h_v \to S$. We want to prove that $g^{-1}(R)$ is dense. But $s\ g \in (a\ F\ h_v, a\ F\ h_u)$ represents some element x in $\lambda\ a\ F\ h_u(v) \subseteq$ $\subseteq a\ \lambda\ a\ F\ h_u(v)$; the commutativity of the diagram shows that this element x is in fact in $a\lambda R(v)$. This implies the existence of a covering $v = \bigvee_{i \in I} v_i$ in H such that for any i, $x|_{v_i}$ is in $\lambda R(v_i)$ and is thus represented by $\overline{g}_i \in (R_i, R)$ with $R_i \in J(v_i)$. Then the following diagram is commutative (proposition 41).

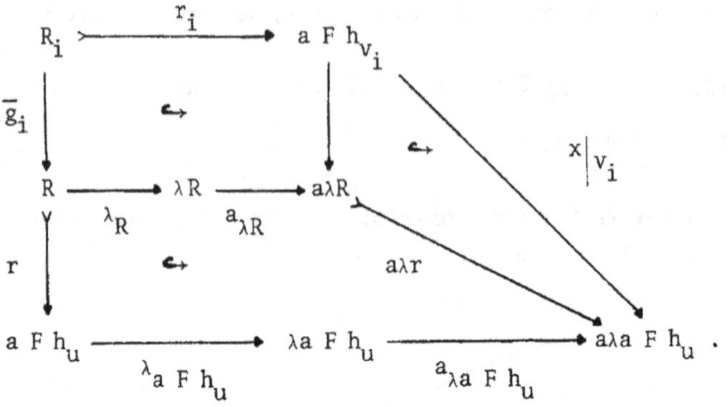

Again by proposition 41, the following diagram is also commutative

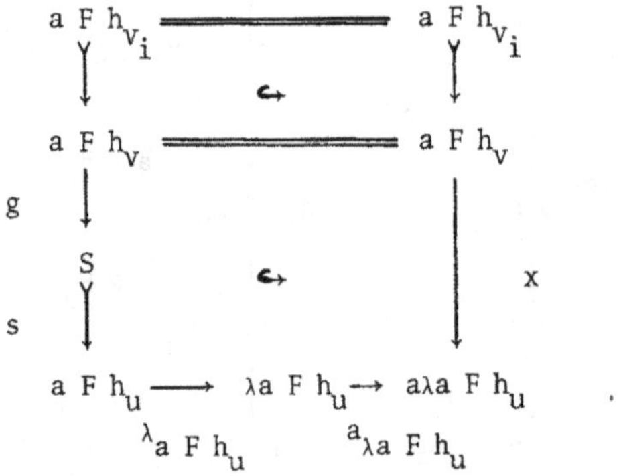

and therefore $s \circ g|_{v_i}$ and $r \circ \overline{g}_i$ represent the same element in $\lambda a\ F\ h_u(v_i)$. Thus they coincide on some $S_i \in J(v_i)$. So, consider the following diagram where the squares are pullbacks

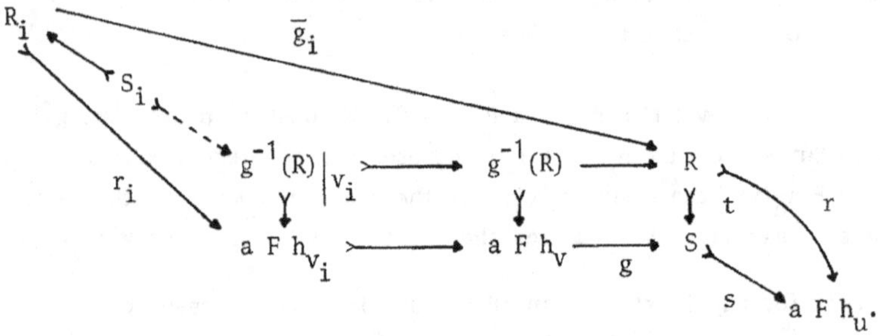

All composites from S_i to $a\ F\ h_u$ are equal and since s is a monomorphism, all composites from S_i to S are equal. So we get a monomorphic factorization

$S_i \rightarrowtail g^{-1}(R)\big|_{v_i}$ which shows, because $S_i \in J(v_i)$, that each $g^{-1}(R)\big|_{v_i}$ is in $J(v_i)$. But this implies that $g^{-1}(R)$ is in $J(v)$ (by (G G 4)). This conclues the proof. ■

Proposition 52.

The correspondence sending a localization of Sh($\mathbb{H}, \mathbb{T}$) *to the corresponding universal closure operation is injective.*

Let $\mathbb{L} \underset{i}{\overset{\ell}{\leftrightarrows}} \mathrm{Sh}(\mathbb{H}, \mathbb{T})$ be a localization. By [21] - 19 - 3 - 1, f, $\mathbb{L} \cong \mathrm{Sh}(\mathbb{H}, \mathbb{T})\ [\Sigma^{-1}]$ where Σ is the class of all morphisms in Sh($\mathbb{H}, \mathbb{T}$) sended by ℓ to an isomorphism. Thus a localization of Sh($\mathbb{H}, \mathbb{T}$) is completely characterized by the class Σ of morphisms f such that $\ell(f)$ is an isomorphism. But, by proposition 48, Σ is itself completely characterized by its monomorphisms. And by proposition 50 the monomorphisms in Σ are completely characterized by the corresponding universal closure operation. Finally, the localization is completely characterized by the corresponding universal closure operation. ■

Proposition 53.

The correspondence sending a universal closure operation on Sh($\mathbb{H}, \mathbb{T}$) *to the corresponding Lawvere - Tierney* $\mathbb{T}$*-topology on* $\mathbb{H}$ *is injective.*

Consider two universal closure operations $R \mapsto \overline{R}$ and $R \mapsto \widetilde{R}$ which give rise to the same Lawvere - Tierney $\mathbb{T}$-topology, i.e. which coincide on the subobjects of each $a\,F\,h_u$, $u \in \mathbb{H}$. Consider any monomorphism $R \rightarrowtail A$ in Sh($\mathbb{H}, \mathbb{T}$) and any $u \in \mathbb{H}$. An element in $\overline{R}(u)$ is represented by a morphism $f : a\,F\,h_u \to A$ which factors through $\overline{R}$; thus the following square is a pullback

$$\begin{array}{ccc} a\,F\,h_u & \xrightarrow[f]{} & \overline{R} \\ \| & & \downarrow \\ a\,F\,h_u & \xrightarrow[f]{} & A. \end{array}$$

But

$$\begin{aligned} a\,F\,h_u &= f^{-1}(\overline{R}) \\ &= \overline{f^{-1}(R)} \\ &= \widetilde{f^{-1}(R)} \\ &= f^{-1}(\widetilde{R}). \end{aligned}$$

This shows that f factors through $\tilde{R}$ and thus $f \in \tilde{R}(u)$. Finally $\bar{R} \leqslant \tilde{R}$. In the same way, $\tilde{R} \leqslant \bar{R}$ and thus $\bar{R} = \tilde{R}$. This concludes the proof. ■

Proposition 54.

The correspondence sending a Lawvere - Tierney $\mathbb{T}$*-topology on* $\mathbb{H}$ *to a Gabriel - Grothendieck* $\mathbb{T}$*-topology on* $\mathbb{H}$ *is injective.*

Suppose that two Lawvere - Tierney $\mathbb{T}$-topologies j and j' give rise to the same Gabriel - Grothendieck $\mathbb{T}$-topology. Consider $u \in \mathbb{H}$ and $R \in \Omega_{\mathbb{T}}(u)$; we must prove that $j(u)(R) = j'(u)(R)$.

Consider any $v \in \mathbb{H}$ and any $f : a\,F\,h_v \to j(u)(R)$, i.e. any element $f \in j(u)(R)(v)$. f is thus an element in $a\,F\,h_u(v)$. By construction of the associated sheaf functor, there is a covering $v = \bigvee_{i \in I} v_i$ in $\mathbb{H}$ such that each $f|_{v_i}$ is in $F\,h_u(v_i)$, i.e. is a morphism $F\,h_{v_i} \to F\,h_u$. Now consider the following pullbacks

$$\begin{array}{ccccc} a\,F\,h_{v_i} & \rightarrowtail & a\,F\,h_v & \xrightarrow{f} & j(u)(R) \\ \| & & \| & & \downarrow \\ a\,F\,h_{v_i} & \rightarrowtail & a\,F\,h_v & \xrightarrow{f} & a\,F\,h_u . \end{array}$$

If $v_i \not\leqslant u$, we know by lemma 9 that

$$f^{-1}(R)\Big|_{v_i} = (f\big|_{v_i})^{-1}(R) = a\,F\,h_{v_i}$$

and in particular :

$$j(v_i)(f^{-1}(R)\Big|_{v_i}) = a\,F\,h_{v_i} .$$

If $v_i \leqslant u$, we know by lemma 8 that $f|_{v_i}$ can be factored in the following way :

$$a\,F\,h_{v_i} \xrightarrow{\alpha} a\,F\,h_{v_i} \rightarrowtail a\,F\,h_u .$$

Since j is a morphism of presheaves, we have :

$$j(v_i)(R\big|_{v_i}) = j(u)(R)\Big|_{v_i}$$

and since j is a morphism in $\&(\mathbb{H}, \mathbb{T})$

$$j(v_i)(\alpha^{-1}(R|_{v_i})) = \alpha^{-1}(j(u)(R)|_{v_i}).$$

Hence,

$$j(v_i)(f^{-1}(R)|_{v_i}) = f^{-1}(j(u)(R))|_{v_i} = a\,F\,h_{v_i}.$$

Thus for any i, $f^{-1}(R)|_{v_i} \in J(v_i)$.

Now look at j'. In the same way, if $v_i \not\leqslant u$, we know by lemma 9 that

$$f^{-1}(j'(u)(R)|_{v_i} = a\,F\,h_{v_i}.$$

Now if $v_i \leqslant u$, we have again by lemma 8

$$f^{-1}(j'(u)(R))|_{v_i} = j'(v_i)(f^{-1}(R)|_{v_i})$$

$$= a\,F\,h_{v_i};$$

the last equality holds because $f^{-1}(R)|_{v_i}$ is in $J(v_i)$. But these relations show that the two elements $f^{-1}(j'(u)(R))|_v$ and $a\,F\,h_v$ of $\Omega_{\mathbb{T}}(v)$ have the same restrictions in each $\Omega_{\mathbb{T}}(v_i)$. So, since $\Omega_{\mathbb{T}}$ is a sheaf, $f^{-1}(j'(u)(R))|_v =$ $= a\,F\,h_v$. In other words, f factors through $j'(u)(R)$ and determines an element in $j'(u)(R)(v)$. Finally, we have shown the inclusion $j(u)(R) \leqslant j'(u)(R)$. Conversely, we have $j'(u)(R) \leqslant j(u)(R)$ and finally $j(u)(R) = j'(u)(R)$. So $j = j'$ and the proposition is proved. ∎

Proposition 55.

The correspondence which sends a Gabriel – Grothendieck $\mathbb{T}$*-topology on* $\mathbb{H}$ *to the corresponding localization of* $Sh(\mathbb{H}, \mathbb{T})$ *is injective.*

Consider two Gabriel - Grothendieck topologies J and J' on $\mathbb{H}$ which correspond to the same localization $\mathbb{L} \underset{i}{\overset{\ell}{\rightleftarrows}} Sh(\mathbb{H}, \mathbb{T})$. For any u in $\mathbb{H}$, $J(u)$ and $J'(u)$ are equal to the set of those subobjects $r : R \rightarrowtail a\,F\,h_u$ such that ℓr is an isomorphism (proposition 51). Thus $J = J'$. ∎

Theorem 56.

The results of § 5 – 6 – 7 – 8 *describe one-to-one correspondences between*

(1) *the localizations of* $Sh(\mathbb{H}, \mathbb{T})$,

(2) *the universal closure operations on* $Sh(\mathbb{H}, \mathbb{T})$,

(3) *the Lawvere - Tierney* $\mathbb{T}$*-topologies on* $\mathbb{H}$,

(4) *the Gabriel - Grothendieck* $\mathbb{T}$*-topologies on* $\mathbb{H}$.

By propositions 52 - 53 - 54 - 55, we have injections

$$\begin{array}{ccc} \{\text{localizations}\} & \overset{\delta}{\longleftarrow\!\!\!+} & \{\text{G - G } \mathbb{T}\text{-topologies}\} \\ \alpha \updownarrow & & \updownarrow \gamma \\ \{\text{closure operations}\} & \underset{\beta}{+\!\!\!\longrightarrow} & \{\text{L - T } \mathbb{T}\text{-topologies}\}. \end{array}$$

Let us prove that $\delta \circ \gamma \circ \beta \circ \alpha = \mathrm{id}$. To that end, consider a localization $\mathbb{L} \underset{i}{\overset{\ell}{\leftrightarrows}} \mathrm{Sh}(\mathbb{H}, \mathbb{T})$. Via α, β, γ, the corresponding Gabriel - Grothendieck $\mathbb{T}$-topology J is given by

$$\begin{aligned} J(u) &= \{R \in \Omega_{\mathbb{T}}(u) \mid \bar{R} = a\,F\,h_u\} \\ &= \{R \in \Omega_{\mathbb{T}}(u) \mid \ell R \cong \ell\, a\,F\,h_u\}; \end{aligned}$$

the last equality holds by proposition 50.

Now consider the localization $\mathbb{L}' \underset{i'}{\overset{\ell'}{\leftrightarrows}} \mathrm{Sh}(\mathbb{H}, \mathbb{T})$ given by $\delta(J)$. By proposition 51, a monomorphism $r : R \rightarrowtail a\,F\,h_u$ is such that $\ell' r$ is an isomorphism if and only if r is in $J(u)$. So, $\ell' r$ is a monomorphism if and only if ℓr is a monomorphism. But we know already (proof of proposition 52) that two localizations are equivalent when they transform the same monomorphisms into isomorphisms. But then, by proposition 49, the only monomorphisms to be considered are those with codomain an object of the form $a\,F\,h_u$. Finally, both localizations $\mathbb{L} \underset{i}{\overset{\ell}{\leftrightarrows}} \mathrm{Sh}(\mathbb{H}, \mathbb{T})$ and $\mathbb{L}' \underset{i'}{\overset{\ell'}{\leftrightarrows}} \mathrm{Sh}(\mathbb{H}, \mathbb{T})$ are equivalent because they transform the same monomorphisms $r : R \rightarrowtail a\,F\,h_u$ into isomorphism.

We know already that $\delta \circ \gamma \circ \beta \circ \alpha = \mathrm{id}$. Thus $\delta \circ \gamma \circ \beta \circ \alpha \circ \delta = \delta$ and since δ is injective, $\gamma \circ \beta \circ \alpha \circ \delta = \mathrm{id}$. In the same way, $\gamma \circ \beta \circ \alpha \circ \delta \circ \gamma = \gamma$ and thus $\beta \circ \alpha \circ \delta \circ \gamma = \mathrm{id}$; $\beta \circ \alpha \circ \delta \circ \gamma \circ \beta = \beta$ and thus $\alpha \circ \delta \circ \gamma \circ \beta = \mathrm{id}$. This concludes the proof. ∎

§ 10. THE CASE OF GROUPS AND ABELIAN GROUPS

In chapter 6, the results of the present chapter (as well as those of chapter 5) will be particularized to the case of a theory of modules on a ring. In this paragraph, we shall treat the more particular case of abelian groups. This investigation will also provide some information about non abelian groups and show why our results fail for the latter. This justifies our assumption that $\mathbb{T}$ is commutative.

We consider the frame $\mathbb{H}$ of open subsets of the singleton, i.e. the initial frame $\{0, 1\}$ and the theory $\mathbb{T}$ of abelian groups. Thus $Sh(\mathbb{H}, \mathbb{T})$ is the category Ab of abelian groups and $Sh(\mathbb{H})$ is the category Sets of sets. The two representable algebras are $a\,F\,h_0 = (o)$ and $a\,F\,h_1 = (\mathbb{Z}, +)$. The monoïd $M_{\mathbb{T}}$ is thus the monoïd of 1-ary operations of $\mathbb{T}$, which is $(\mathbb{Z}, \times)$. So the topos $\mathcal{E}(\mathbb{H}, \mathbb{T})$ is the topos of $(\mathbb{Z}, \times)$-sets.

$\Omega_{\mathbb{T}}$ is the set of subgroups of $(\mathbb{Z}, +)$; it is thus isomorphic to $\mathbb{N}$. The action of $(\mathbb{Z}, \times)$ on $\mathbb{N}$ goes as follows : consider $z \in \mathbb{Z}$ and $n \in \mathbb{N}$; z corresponds to the homomorphism $\overline{g} : (\mathbb{Z}, +) \to (\mathbb{Z}, +)$ which is the multiplication by z; n corresponds to the subgroup $(n\mathbb{Z}, +)$.
Therefore the action

$$(\mathbb{Z}, \times) \times \mathbb{N} \to \mathbb{N};\ (z, n) \mapsto z * n$$

is defined by the fact that $z * n$ is the generator of $\overline{z}^{-1}(n\ \mathbb{Z})$. But

$$\begin{aligned} z^{-1}(n\ \mathbb{Z}) &= \{x \in \mathbb{Z} \mid z\,x \in n\ \mathbb{Z}\} \\ &= \{x \in \mathbb{Z} \mid n \text{ divides } z\,x\} \\ &= \{x \in \mathbb{Z} \mid \tfrac{n}{n \wedge z} \text{ divides } x\} \\ &= \frac{n}{n \wedge z}\ \mathbb{Z} \end{aligned}$$

where $n \wedge z$ denotes the greatest common divisor of n and z. Thus $z * n = \frac{n}{n \wedge z}$.

$\mathbb{N} = \Omega_{\mathbb{T}}$ has an ordering corresponding to the inclusion $n\ \mathbb{Z} \leqslant m\ \mathbb{Z}$ of subgroups; thus $n \triangleleft m$ in $\mathbb{N} = \Omega_{\mathbb{T}}$ if and only if m divides n. The greatest element is 1 and the intersection $n \vartriangle m$ corresponds to the intersection $n\ \mathbb{Z} \cap m\ \mathbb{Z}$ of subgroups; thus $n \vartriangle m = n \vee m$ where $n \vee m$ is the smallest common multiple of n and m.

Now consider some subgroup $R \rightarrowtail A$ in Ab. There is a characteristic map $\varphi : A \to \mathbb{N}$. If $a \in A$, denote $\overline{a} : \mathbb{Z} \to A$ the homomorphism which sends 1 to a. $\overline{a}^{-1}(R)$ is some subgroup of $\mathbb{Z}$, thus has the form $n\ \mathbb{Z}$; n is the value of $\varphi(a)$. So

$$\overline{a}^{-1}(R) = \{z \in \mathbb{Z} \mid a\,z \in \mathbb{R}\}$$

and thus

$$\varphi(a) = \min\{n \in \mathbb{N}^* \mid n\,a \in R\}$$

if this set is non empty and $\varphi(a) = 0$ if this set is empty. (! The minimum is computed for the usual ordering on $\mathbb{N}$!).

If A is any abelian group, a map $\varphi : A \to \mathbb{N}$ is a characteristic map if it is a morphism in $\mathcal{E}(\mathbb{H}, \mathbb{T})$ satisfying the conditions of definition 19. φ is a morphism in $\mathcal{E}(\mathbb{H}, \mathbb{T})$ if and only if for any $a \in A$ and $z \in A$, $\varphi(za) = z * \varphi(a)$, that is to say

$$\varphi(za) = \frac{\varphi(a)}{z \wedge \varphi(a)}.$$

Now φ must satisfy a condition for any operation of the theory; it suffices to express these conditions for the basic operations 0, +, -. From the remark following definition 19, the condition on 0 is

$$\varphi(0) = 1.$$

The condition on + means

$$\varphi(a + b) \rhd (\varphi(a) \vartriangle \varphi(b))$$

that is to say

$$\varphi(a + b) \text{ divides } \varphi(a) \vee \varphi(b).$$

The condition on - means

$$\varphi(-a) \rhd \varphi(a)$$

that is to say

$$\varphi(-a) \text{ divides } \varphi(a)$$

which is already satisfied because

$$\varphi(-a) = \frac{\varphi(a)}{(-1) \wedge \varphi(a)} = \varphi(a).$$

To summarize, a mapping $\varphi : A \to \mathbb{N}$ is the characteristic mapping of some subgroup of A if and only if

(1) $\varphi(0) = 1$

(2) $\varphi(za) = \dfrac{\varphi(a)}{z \wedge \varphi(a)}$ $\qquad z \in \mathbb{Z}$

(3) $\varphi(a + b)$ divides $\varphi(a) \vee \varphi(b)$.

A Lawvere - Tierney $\mathbb{T}$-topology on $\mathbb{N}$ is a mapping $j : \mathbb{N} \to \mathbb{N}$ which is compatible with the action of $(\mathbb{Z}, \times)$ and satisfies (LT -1 - 2 - 3). This means

(1) $j(\dfrac{n}{n \wedge z}) = \dfrac{j(n)}{j(n) \wedge z}$; $\qquad z \in \mathbb{Z}$

(2) $j(1) = 1$

(3) $j\, j(n) = j(n)$

(4) $j(n \vee m) = jn \vee jm$.

If p is some prime number, the usual localization process at the prime ideal $p\mathbb{Z}$ is thus described by such a morphism $j : \mathbb{N} \to \mathbb{N}$. If $n\mathbb{Z} \hookrightarrow \mathbb{Z}$ is some ideal of $\mathbb{Z}$, its localization is obtained by adding an element $\frac{1}{a}$ for any $a \in n\mathbb{Z}$ which cannot be divided by p. Now consider $n\mathbb{Z} \subseteq m\mathbb{Z}$, i.e. m divides n. These two ideals have the same localization if and only if each element of the form $\frac{m\ a}{m\ b}$ where p does not divide m b can be written in the form $\frac{n\ a}{n\ b}$ where p does not divide n b. But clearly, this is possible if and only if $\frac{n}{m}$ cannot be divided by p. Thus the greatest ideal $m\mathbb{Z}$ which has the same localization as $n\mathbb{Z}$ obtained when m is the smallest integer such that $\frac{n}{m}$ cannot be divided by p, i.e. when m is the greatest power of p dividing n. But the greatest ideal which has the same localization as $n\mathbb{Z}$ is the closure of $n\mathbb{Z}$ for the universal closure operation associated to the localization (via proposition 50). Thus the Lawvere - Tierney $\mathbb{T}$-topology corresponding to the localization associated to some prime number p is given by

$$j : \mathbb{N} \to \mathbb{N}$$
$$j(n) = \text{greatest power of p dividing n.}$$

Consider now the theory $\mathbb{T}$ of arbitrary groups. Again, let $\mathbb{H}$ be the initial frame. Then $Sh(\mathbb{H}, \mathbb{T})$ is the category Gr of groups. The two representable functors are again $a\ F\ h_0 = (0)$ and $a\ F\ h_1 = (\mathbb{Z}, +)$. Thus the monoïd $M_{\mathbb{T}}$ of 1-ary operations, the topos $\mathcal{E}(\mathbb{H}, \mathbb{T})$ and the object $\Omega_{\mathbb{T}}$ can be constructed in the same way and are exactly the same as in the commutative case. But the principal results of this chapter are no longer valid and this explains why we required $\mathbb{T}$ to be commutative. We will show that theorems 22 and 56 fail in the case of groups.

Consider the coproduct $\mathbb{Z} \perp\!\!\!\perp \mathbb{Z}$ in Gr, where the two basic generators are denoted by x and y. Consider the subgroup R generated by the words xx and yy. The corresponding characteristic mapping $\varphi : \mathbb{Z} \perp\!\!\!\perp \mathbb{Z} \to \mathbb{N}$ is thus given by

$$\varphi(w) = \begin{cases} \inf \{n \mid n\ w \in R\} \\ \text{or } 0 \text{ if this set is empty.} \end{cases}$$

In particular, $\varphi(x) = z$, $\varphi(y) = z$, $\varphi(xy) = 0$. Thus the rule $\varphi(xy)$ divides $\varphi(x) \vee \varphi(y)$ is not satisfied in this case.

In order to compare localizations and $\mathbb{T}$-topologies, observe that Ab and Gr have the same representable objects and thus the same Gabriel - Grothendieck $\mathbb{T}$-topologies; they have also the same topos $\mathcal{E}(\mathbb{H}, \mathbb{T})$ and the same object $\Omega_{\mathbb{T}}$, thus the same Lawvere - Tierney $\mathbb{T}$-topologies. For Ab, these $\mathbb{T}$-topologies are in one-to-one correspondance with the localizations of Ab (theorem 56) and there are many non obvious such localizations (see [6], [20] or even our easier example of localizing at some prime number p). On the other hand, the category Gr of groups has only the two trivial localizations (0) and Gr (proposition 57). Thus theorem 56 fails in the case of the category Gr of groups.

Proposition 57.

(0) and Gr are the only two localizations of the category Gr of groups.

Consider a localization $\mathbb{L} \underset{i}{\overset{\ell}{\leftrightarrows}} \mathrm{Gr}$ of Gr which is not the identity on Gr. Observe that propositions 48 - 49 - 50 - 52 do not depend on the commutativity of $\mathbb{T}$. ℓ is not the identity thus ℓ takes some non-isomorphic morphism into an isomorphism (proof of proposition 52) and thus finally some $s : n\mathbb{Z} \hookrightarrow \mathbb{Z}$ $(n \neq 1)$ into an isomorphism (propositions 48 - 49).

Denote by f the morphism which sends 1 to the word $x\,y \in \mathbb{Z} \amalg \mathbb{Z}$, where x and y are the two basic generators of $\mathbb{Z} \amalg \mathbb{Z}$. Consider the following square which is obviously a pullback (because $n \neq 1$).

$$\begin{array}{ccc} (0) & \longrightarrow & n\mathbb{Z} \amalg n\mathbb{Z} \\ {\scriptstyle r}\big\downarrow & & \big\downarrow{\scriptstyle s \amalg s} \\ \mathbb{Z} & \xrightarrow[f]{} & \mathbb{Z} \amalg \mathbb{Z} \end{array}$$

$\ell(s)$ is an isomorphism, thus $\ell(s \amalg s)$ is an isomorphism and $\ell(r)$ is an isomorphism (ℓ is exact). Thus $\ell(\mathbb{Z}) \cong \ell(0)$. But ℓ is exact and therefore $\ell(\mathbb{Z}) \cong (0)$. Now $\mathbb{Z}$ is a generator and any group can be reconstructed as a colimit of a diagram containing only $\mathbb{Z}$; since ℓ preserves colimits, $\ell(G) \cong (0)$ for any group G. Thus the localization is just (0). ■

For the reader who is interested in the non commutative case, let us point out that if $\mathbb{L} \underset{i}{\overset{\ell}{\leftrightarrows}} \mathrm{Sh}(\mathbb{H}, \mathbb{T})$ is a localization of $\mathrm{Sh}(\mathbb{H}, \mathbb{T})$ for an arbitrary theory $\mathbb{T}$, the sets defined by

$$J(u) = \{ r : R \rightarrowtail a\,F\,h_u \mid \ell(r) \text{ is iso} \}$$

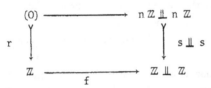

for any $u \in \mathbb{H}$, describe a Gabriel - Grothendieck $\mathbb{T}$-topology on $\mathbb{H}$ in the sense of definition 30. Moreover, it is still true that this correspondence from a localization to a $\mathbb{T}$-topology is injective. Thus any localization of Sh($\mathbb{H}$, $\mathbb{T}$) is completely characterized by a $\mathbb{T}$-topology on $\mathbb{H}$, but there are $\mathbb{T}$-topologies on $\mathbb{H}$ which do not derive from a localization of Sh($\mathbb{H}$, $\mathbb{T}$) - (counterexample of groups). So the localizations of Sh($\mathbb{H}$, $\mathbb{T}$) are exactly classified by $\mathbb{T}$-topologies on $\mathbb{H}$ satisfying additional axioms. To conclude, the notion of $\mathbb{T}$-topology does not seem to be the right one in the non commutative case. Or maybe the usual notion of localization is not the right one in the non commutative case. This problem is open.

CHAPTER 4 : INTEGRAL THEORIES AND CHARACTERIZATION THEOREM

Let $\mathbb{H}$ be a frame and $Sh(\mathbb{H})$ the topos of sheaves over $\mathbb{H}$. It is well-known that $\mathbb{H}$ can be recovered from $Sh(\mathbb{H})$. $\mathbb{H}$ is exactly the frame of subobjects of 1 in $Sh(\mathbb{H})$.

Now consider a fixed external theory $\mathbb{T}$. We ask an analogous question : can $\mathbb{H}$ be recovered from $Sh(\mathbb{H}, \mathbb{T})$? The answer is "no" in general : in § 1, we provide a counterexample. We consider the boolean ring $2^{\mathbb{N}}$ (power set of $\mathbb{N}$) and the corresponding (external) theory $\mathbb{T}$ of $2^{\mathbb{N}}$-modules. We prove that for any non zero integers n, m, $Sh(2^n, \mathbb{T})$ is equivalent to $Sh(2^m, \mathbb{T})$.

However, if we add some assumption on the theory $\mathbb{T}$, the answer becomes "yes". The assumption is that each non constant 1-ary operation is an epimorphism in $\mathbb{T}$, where we regard $\mathbb{T}$ as a category with a denumerable set of objects. In the case of the theory $\mathbb{T}$ of modules on the ring R, this is equivalent to the fact that R is an integral domain. Such theories are therefore called integral. The notion of integral theory includes as examples the theories of sets, monoïds, groups, rings, modules on an integral domain (in particular abelian groups and vector spaces), algebras on a field, boolean algebras, sets on which a group acts, and so on ... . Integral theories are studied in § 2.

In § 3, we prove the characterization theorem. If $\mathbb{T}$ is an integral theory, two frames $\mathbb{H}$ and $\mathbb{H}'$ are isomorphic if and only if the categories $Sh(\mathbb{H}, \mathbb{T})$ and $Sh(\mathbb{H}', \mathbb{T})$ are equivalent. To prove this, we show that for an integral theory $\mathbb{T}$, $\mathbb{H}$ is just the frame of formal initial segments of $Sh(\mathbb{H}, \mathbb{T})$. In particular, $\mathbb{H}$ can be recovered from $Sh(\mathbb{H}, \mathbb{T})$.

§ 1. A COUNTEREXAMPLE

Consider an algebraic theory $\mathbb{T}$ in the usual sense. For any frame $\mathbb{H}$, $\mathbb{T}$ can be seen as a theory in $Pr(\mathbb{H})$ (= the constant functor $\mathbb{T}$) or in $Sh(\mathbb{H})$ (= the corresponding associated sheaf functor). Thus when $\mathbb{T}$ is fixed externally, $Sh(\mathbb{H}, \mathbb{T})$ depends only on $\mathbb{H}$. So it makes sense to ask if $Sh(\mathbb{H}, \mathbb{T})$ characterizes $\mathbb{H}$ completely. In this paragraph, we exhibit a counterexample.

Consider the set $\mathbb{N}$ of integers (or more generally any infinite set E). Choose any bijection $\mathbb{N} \cong \mathbb{N} \amalg \mathbb{N}$ and apply the power set functor, which takes colimits into limits (it is represented by 2)

$$2^{\mathbb{N}} \cong 2^{\mathbb{N} \amalg \mathbb{N}} \cong 2^{\mathbb{N}} \times 2^{\mathbb{N}}.$$

This isomorphism of $2^{\mathbb{N}}$ and $2^{\mathbb{N}} \times 2^{\mathbb{N}}$ is in fact an isomorphism of boolean rings (the addition is the symmetric difference $A \Delta B = (A \cup B) \setminus (A \cap B)$ and the multiplication is the intersection $A \cap B$), (see [11] - II - 3 - 9).

From the isomorphism of rings $2^{\mathbb{N}} \cong 2^{\mathbb{N}} \times 2^{\mathbb{N}}$, we deduce an equivalence between the corresponding categories of modules

$$\underline{\mathrm{Mod}}_{2^{\mathbb{N}}} \cong \underline{\mathrm{Mod}}_{2^{\mathbb{N}} \times 2^{\mathbb{N}}} \cong \underline{\mathrm{Mod}}_{2^{\mathbb{N}}} \times \underline{\mathrm{Mod}}_{2^{\mathbb{N}}}$$

(cfr. [17]). But $\underline{\mathrm{Mod}}_{2^{\mathbb{N}}}$ is the category of sheaves of $2^{\mathbb{N}}$-modules on the one-point topological space and $\underline{\mathrm{Mod}}_{2^{\mathbb{N}}} \times \underline{\mathrm{Mod}}_{2^{\mathbb{N}}}$ is the category of sheaves of $2^{\mathbb{N}}$-modules on the discrete two points set. Thus we have proved the equivalence

$$\mathrm{Sh}(2^1, \mathbb{T}) \cong \mathrm{Sh}(2^2, \mathbb{T})$$

where $\mathbb{T}$ is the theory of $2^{\mathbb{N}}$-modules. Iterating the process, we deduce that for any two non zero integers n, m

$$\mathrm{Sh}(2^n, \mathbb{T}) \cong \mathrm{Sh}(2^m, \mathbb{T}).$$

This is clearly a counterexample to our problem since for $n \neq m$, 2^n and 2^m are not isomorphic.

§ 2. INTEGRAL THEORIES

If we consider a category $\mathrm{Sh}(\mathbb{H}, \mathbb{T})$, where $\mathbb{T}$ is a theory and $\mathbb{H}$ a frame, the objects $a\, F\, h_u$ generate $\mathrm{Sh}(\mathbb{H}, \mathbb{T})$, (proposition I - 6). But any hom-set in $\mathbb{H}$ is reduced either to a singleton or to the empty set : therefore the generators $a\, F\, h_u$ are constructed from the two free algebras F0 and F1. This explains why assumptions on 1-ary operations (= elements of F1) and constants (= elements of F0) of $\mathbb{T}$ induce strong consequences on $\mathrm{Sh}(\mathbb{H}, \mathbb{T})$, even without any assumption on n-ary operations with $n \geq 1$. In this paragraph, we describe such assumptions : integrability of the theory with, as special cases, left or right simplicity.

As mentioned in the introduction of this chapter, a theory $\mathbb{T}$ is integral if any non constant 1-ary operation is an epimorphism in $\mathbb{T}$ or, equivalently, if any non constant $\mathbb{T}$-endomorphism of F1 is injective. This is in particular the case when $\mathbb{T}$ is left simple, i.e. when F1 has only the trivial subobjects

F0 and F1, or when $\mathbb{T}$ is right simple, i.e. when F1 has only the trivial quotients F0 (if there are constants) and F1. Left simplicity is also equivalent to the fact that any non constant 1-ary operation has a section.

As already mentioned, for a theory $\mathbb{T}$ of modules on a ring R, integrability of $\mathbb{T}$ is equivalent to the fact that R is an integral domain. Other examples are given by sets, monoïds, groups, rings, boolean algebras, and so on ... .

We start with two technical lemmas which make explicit and precise the form of some well-known isomorphisms.

<u>Lemma 1.</u>

For a theory $\mathbb{T}$, *there is a contravariant homomorphism of monoïds (for the usual composition of arrows) :*

$$\mathbb{T}(T^1, T^1) \cong \underline{Sets}^{\mathbb{T}}(F1, F1).$$

Looking at an algebra as a product preserving functor $\mathbb{T} \to$ Sets, F1 is just $\mathbb{T}(T^1, -)$, (cfr. [21] - 18). Thus by the Yoneda lemma, we have $\underline{Sets}^{\mathbb{T}}(F1, F1) \underset{def}{=} Nat(\mathbb{T}(T^1, -), \mathbb{T}(T^1, -)) \cong \mathbb{T}(T^1, T^1)$ and the last isomorphism takes $\mathbb{T}(\alpha, -)$ to $\alpha : T^1 \to T^1$. From the equality

$$\mathbb{T}(\alpha, -) \circ \mathbb{T}(\beta, -) = \mathbb{T}(\beta \circ \alpha, -)$$

we deduce that the isomorphism is contravariant. ■

<u>Lemma 2.</u>

For a theory $\mathbb{T}$, *the bijections*

$$\mathbb{T}(T^1, T^1) \cong \underline{Sets}^{\mathbb{T}}(F1, F1) \cong F1.$$

$$\alpha \longleftrightarrow \overline{\alpha} \longleftrightarrow \alpha$$

are such that $\overline{\alpha}(\widetilde{\beta}) = \widetilde{\beta \circ \alpha}$.

The first bijection is the one given by proposition 1 and the second comes from the adjunction $F \dashv U$ where $U : \underline{Sets}^{\mathbb{T}} \to \underline{Sets}$ is the forgetful functor :

$$\underline{Sets}^{\mathbb{T}}(F1, F1) \cong \underline{Sets}(1, UF1) \cong UF1.$$

F1, seen as a product preserving functor $\mathbb{T} \to \underline{Sets}$ is just $\mathbb{T}(T^1, -)$ and $\overline{\alpha}$

is just $\mathbb{T}(\alpha, -)$, (cfr. proposition 1). So for any integer n, $\mathbb{T}(\alpha, T^n)$: $\mathbb{T}(T^1, T^n) \to \mathbb{T}(T^1, T^n)$ is defined by $\mathbb{T}(\alpha, T^n)(\gamma) = \gamma \circ \alpha$. In particular $\mathbb{T}(\alpha, T^1) : \mathbb{T}(T^1, T^1) \to \mathbb{T}(T^1, T^1)$ is defined by $\mathbb{T}(\alpha, T^1)(\beta) = \beta \circ \alpha$. But $\mathbb{T}(\alpha, T^1) \cong \overline{\alpha}$ and thus $\overline{\alpha}(\widetilde{\beta}) = \widetilde{\beta \circ \alpha}$. ∎

Proposition 3.

For a theory $\mathbb{T}$, *a 1-ary operation* $\alpha : T^1 \to T^1$ *is an epimorphism if and only if for any 1-ary operations* $\beta, \gamma : T^1 \rightrightarrows T^1$

$$\beta \circ \alpha = \gamma \circ \alpha \quad \Rightarrow \quad \beta = \gamma.$$

If α is an epimorphism, the condition certainly holds. Now suppose the condition holds and consider $\delta, \varepsilon : T^1 \rightrightarrows T^n$ such that $\delta \circ \alpha = \varepsilon \circ \alpha$. For any projection $p_i : T^n \to T^1$, we then have $p_i \circ \delta \circ \alpha = p_i \circ \varepsilon \circ \alpha$ and since $p_i \circ \delta$ and $p_i \circ \varepsilon$ are 1-ary operations, our assumptions imply $p_i \circ \delta = p_i \circ \varepsilon$. This is true for any i and thus $\delta = \varepsilon$, which proves that α is an epimorphism. ∎

Definition 4.

A theory $\mathbb{T}$ *is integral if and only if any non constant 1-ary operation of* $\mathbb{T}$ *is an epimorphism in* $\mathbb{T}$.

Proposition 5.

A theory $\mathbb{T}$ *is integral if and only if for any non constant 1-ary operation, the corresponding endomorphism of* F1 *is injective.*

By proposition 2, a 1-ary operation α is an epimorphism if and only if for any 1-ary operations β, γ,

$$\beta \circ \alpha = \gamma \circ \alpha \quad \Rightarrow \quad \beta = \gamma.$$

If we denote by $\overline{\alpha}, \overline{\beta}, \overline{\gamma}$ the corresponding endomorphisms of F1, this condition is equivalent to : (proposition 1)

$$\overline{\alpha} \circ \overline{\beta} = \overline{\alpha} \circ \overline{\gamma} \quad \Rightarrow \quad \overline{\beta} = \overline{\gamma}.$$

But F1 is a generator in $\underline{\text{Sets}}^{\mathbb{T}}$; thus this last condition is equivalent to the injectivity of $\overline{\alpha}$. ∎

Proposition 6.

Let $\mathbb{T}$ *be the theory of modules over a fixed ring* R *(with unit).* $\mathbb{T}$ *is integral if and only if* R *is an integral domain.*

The monoïd of 1-ary operations of $\mathbb{T}$ is exactly the monoïd $(R, \times)$. Thus the theory is integral if and only if for any $x \neq 0$ and any y, z

$$y\,x = z\,x \Rightarrow y = z$$

or equivalently

$$(y-z)x = 0 \Rightarrow y - z = 0.$$

Because y, z are arbitrary, this is equivalent to

$$w\,x = 0 \Rightarrow w = 0$$

for $x \neq 0$ and an arbitrary w. In other words, the condition is equivalent to

$$w\,x = 0 \Rightarrow x = 0 \text{ or } w = 0$$

which is the definition of an integral domain. ■

The free algebra functor $F : \underline{\text{Sets}} \to \underline{\text{Sets}}^{\mathbb{T}}$ preserves monomorphisms (cfr. I - 1). So the inclusion $0 \rightarrowtail 1$ is sended to an injection $F0 \rightarrowtail F1$ in $\underline{\text{Sets}}^{\mathbb{T}}$.

<u>Definition 7.</u>

A theory $\mathbb{T}$ *is called left simple if the only subobjects of* F1 *are* F0 *and* F1.

<u>Proposition 8.</u>

A theory $\mathbb{T}$ *is left simple if and only if any non constant 1-ary operation has a section in* $\mathbb{T}$.

Consider a left simple theory $\mathbb{T}$ and a non constant 1-ary operation α. α corresponds to an endomorphism $\overline{\alpha} : F1 \to F1$ and thus to the element $\overline{\alpha}(*) \in F1$ where $*$ is the universal generator of F1. α is non constant, thus $\overline{\alpha}(*)$ is not a constant and the subalgebra generated by $\overline{\alpha}(*)$ must be F1, because $\mathbb{T}$ is left simple. In particular, there is an operation β (necessarily a 1-ary operation) such that $\beta(\overline{\alpha}(*)) = *$. But (with the notations of proposition 2)

$$* = \beta(\overline{\alpha}(*)) = \overline{\alpha}(\beta(*)) = \overline{\alpha}(\widetilde{\beta}) = \widetilde{\beta \circ \alpha}$$

and by proposition 1, this means $\alpha \circ \beta = \mathrm{id}$. Thus α has section β.

Now suppose that any non constant 1-ary operation has a section in $\mathbb{T}$. Take a subalgebra $R \rightarrowtail F1$ which contains a non-constant 1-ary operation $\widetilde{\alpha}$. Choose a 1-ary operation β such that $\alpha \circ \beta = \mathrm{id}$ in $\mathbb{T}$. In F1, we have

$$* = \widetilde{\alpha \circ \beta} = \overline{\alpha}(\widetilde{\beta}) = \overline{\alpha}(\beta(*)) = \beta(\overline{\alpha}(*)) = \beta(\widetilde{\alpha}).$$

But $\widetilde{\alpha} \in R$, thus $\beta(\widetilde{\alpha}) \in R$ and $* \in R$. This proves that $R = F1$. ■

Proposition 9.

Any left simple theory is integral.

Trivial since a morphism with a section is an epimorphism. ■

For any constant $\theta \in F0$, there is a unique homomorphism $\overline{\theta} : F1 \to F0$ such that $\overline{\theta}(*) = \theta$. This homomorphism is surjective because $F0$ is an initial object in $\underline{\text{Sets}}^{\mathbb{T}}$ (F preserves colimits) and thus the composite

$$F0 \rightarrowtail F1 \xrightarrow{\overline{\theta}} F0$$

is the identity on $F0$. Thus $\overline{\theta}$ is a coequalizer (cfr. [21] - 18 - 8 - 8) and $F0$ is thus a quotient of $F1$ via $\overline{\theta}$.

Definition 10.

A theory $\mathbb{T}$ is called right simple if any quotient of $F1$ *is the identity on* $F1$ *or some* $\overline{\theta} : F1 \to F0$.

Proposition 11.

A right simple theory is integral.

Consider a right simple theory $\mathbb{T}$, a non constant 1-ary operation α and two 1-ary operations β, γ such that $\beta \circ \alpha = \gamma \circ \alpha$. We must prove the equality $\beta = \gamma$. Via proposition 1, we consider the following diagram where Q is the coequalizer of $\overline{\beta}$, $\overline{\gamma}$

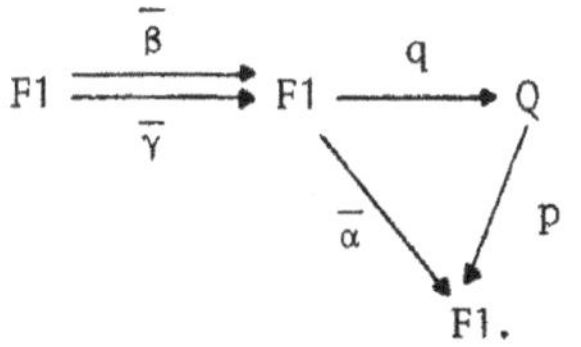

By proposition 1, $\overline{\alpha} \circ \overline{\beta} = \overline{\alpha} \circ \overline{\gamma}$ and thus there is a unique factorization $p : Q \to F1$ such that $p \circ q = \overline{\alpha}$. But the theory is right simple. If $Q = F0$ and $q = \overline{\theta}$ with $\theta \in F0$, then

$$\tilde{\alpha} = \overline{\alpha}(*) = p\ q(*) = p(\theta) = \theta$$

which contradicts the fact that α is non-constant. So q is the identity on F1 and $\overline{\beta} = \overline{\gamma}$, thus $\beta = \gamma$. ■

There are many examples of integral theories.

Example 12 : Sets.

F1 is the singleton which has only two subobjects : $F0 = \phi$ and F1. So the theroy is left simple and thus integral.

Example 13 : Monoïds.

F1 is $(\mathbb{N}, +)$. For $n \in \mathbb{N}$, the homomorphism $\overline{n} : \mathbb{N} \to \mathbb{N}$ is just the multiplication by n. Indeed,

$\overline{n}(0) = 0$

$\overline{n}(1) = \overline{n}(*) = n$

$\overline{n}(m) = \overline{n}(1 + \dots + 1) = \overline{n}(1) + \dots + \overline{n}(1) = n + \dots + n = m\, n.$

If $n \neq 0$, $\overline{n}$ is injective. Thus the theory of monoïds is integral.

Example 14 : Groups.

F1 is $(\mathbb{Z}, +)$. For $z \in \mathbb{Z}$, the homomorphism $\overline{z} : \mathbb{Z} \to \mathbb{Z}$ is just the multiplication by z, (same proof as in example 13, with the additional equality $\overline{z}(-m) = -\overline{z}(m)$). If $z \neq 0$, $\overline{z}$ is injective. Thus the theory of groups is integral.

Example 15 : Abelian groups.

This theory is integral by proposition 6.

Example 16 : Rings.

We do not suppose commutativity nor existence of a unit. F1 is the ring of polynomials

$$a_n X^n + \dots + a_1 X^1$$

where a_i is an integer. If P(X) is any such polynomial, $\overline{P} : F1 \to F1$ takes $X = *$ to P(X) and thus Q(X) on Q(P(X)). Now if

$$Q_1(P(X)) = Q_2(P(X))$$

where P(X) is not 0, we must prove that Q_1 and Q_2 coincide. Suppose P(X) given by

$$P(X) = a_n X^n + \dots + a_1 X^1 \qquad a_n \neq 0.$$

If $Q_1(X)$ has degree m, $Q_1\, P(X)$ has degree $n\, m_1$ and thus also $Q_2(P(X))$; so Q_2 too has degree m. Then, suppose $Q_1(X)$ and $Q_2(X)$ given by

$$Q_1(X) = \alpha_m X^m + \dots + \alpha_1 X^1$$

$$Q_2(X) = \beta_m X^m + \ldots + \beta_1 X^1.$$

We now compare the polynomials $Q_1(P(X)) = Q_2(P(X))$. The terms of degree n m are equal, thus

$$\alpha_m a_n^m = \beta_m a_n^m$$

which implies $\alpha_m = \beta_m$, since $a_n \neq 0$. The terms of degree $n(m-1)$ are equal, thus

$$\alpha_{m-1} a_n^{m-1} + E(\alpha_m, a_n, \ldots, a_1) = \beta_{m-1} a_n^{m-1} + E(\beta_m, a_n, \ldots, a_1)$$

where $E(\alpha_m, a_n, \ldots, a_1)$ is the coefficient of $X^{n(m-1)}$ in $\alpha_m(P(X))^m$. Because $\alpha_m = \beta_m$, we deduce $E(\alpha_m, a_n, \ldots, a_1) = E(\beta_m, a_n, \ldots, a_1)$ and finally $\alpha_{m-1} = \beta_{m-1}$, since $a_n \neq 0$. We iterate the process with the terms of degree n k $(k \leqslant m)$ and finally we prove the equality $\alpha_k = \beta_k$ for all k. Thus $Q_1 = Q_2$ and $\overline{P}$ is injective. So the theory of rings is integral.

Example 17 : Rings with unit.

We do not assume commutativity. Now F1 is the ring $\mathbb{Z}[X]$ and the proof of the integrality of the theory is analogous to that of example 16.

Example 18 : Commutative rings.

We do not assume the existence of a unit. Here F1 is the ring described in example 16 and the proof of the integrality is the same as in example 16.

Example 19 : Commutative rings with unit.

Again F1 is $\mathbb{Z}[X]$ and the proof of the integrality is analogous to that of example 16.

Example 20 : Commutative algebras.

If R is a commutative ring with unit which is an integral domain, the corresponding free commutative algebra with unit F1 is just R[X]. If we do not assume the existence of a unit in the algebra, F1 reduces to those polynomials of the form

$$a_n X^n + \ldots + a_1 X^1.$$

The proof given in example 16 transposes to the present case because R is an integral domain.

Exemple 21 : Modules on an integral domain.

By proposition 6, this theory is integral.

Example 22 : Vector spaces.

Any (skew) field is an integral domain. Thus a theory of vector spaces is integral by proposition 6. The free algebra F1 is the field K itself; it has only the trivial subobjects and quotients : thus the theory is in fact left simple and right simple.

Example 23 : Sets with base point(s).

The free algebra F1 is the set with a single element which is not a base point. The only subobjects are F1 itself and the subset of base points. So the theory is left simple and thus integral.

Example 24 : G-sets for a group G.

The free algebra F1 is the group G itself where the action is the multiplication of the group. If $g \in G$, the homomorphism $\overline{g} : F1 \to F1$ is thus the multiplication by g, which is injective because it has an inverse $\overline{g^{-1}}$. So the theory is integral.

Example 25 : Boolean algebras.

The free algebra F1 is 2^2 :

$$\begin{array}{ccc} & 1 & \\ x & & \complement x \\ & 0 & \end{array}$$

Any subalgebra contains 0 and 1 and if it contains x (resp. $\complement x$) it must contain its complement which is $\complement x$ (resp. x). So the theory of boolean algebras is left simple and thus integral.

§ 3. THE CHARACTERIZATION THEOREM

We now proceed to prove our characterization theorem for an external integral theory $\mathbb{T}$. For any frame $\mathbb{H}$, we prove that $\mathbb{H}$ is isomorphic to the frame of formal initial segments of $Sh(\mathbb{H}, \mathbb{T})$. We deduce that two frames $\mathbb{H}$ and $\mathbb{H}'$ are isomorphic if and only if the categories $Sh(\mathbb{H}, \mathbb{T})$ and $Sh(\mathbb{H}', \mathbb{T})$ are equivalent.

Proposition 26.

Let $\mathbb{H}$ be a frame and $\mathbb{T}$ an external integral theory. The canonical inclusion *of* $\mathbb{H}$ *in the frame* $\mathcal{H}$ *of formal initial segments of* $\mathrm{Sh}(\mathbb{H}, \mathbb{T})$ *is an isomorphism of frames.*

Up to now, we have described in the following way the internal theory $\mathbb{T}$ associated to the external theory $\mathbb{T}$: it is the sheaf associated to the constant presheaf $\mathbb{T}$ (external). But this constant presheaf is generally not a separated presheaf : indeed for a separated presheaf P, P(o) has at most one element (o is covered by the empty covering and thus two elements of P(o) have always the same restriction at all the (unexisting) elements of the empty covering; thus they must be equal). In fact, this difficulty at o is the only reason why a constant presheaf Q is generally not a sheaf. Indeed, if $u \neq o$ in $\mathbb{H}$ and $u = \bigvee_{i \in I} u_i$, at least one u_i is distinct from o; say $u_j \neq o$. Thus if $x, y \in Q(u)$, and for any $i \in I$, $x|_{u_i} = y|_{u_i}$, then in particular

$$x = x|_{u_j} = y|_{u_j} = y.$$

On the other hand, if A is a sheaf on $\mathbb{H}$, then A(o) is exactly a singleton. A is separated, thus we know already that A(o) has at most an element. Now the empty family of elements is a compatible family for the empty covering of o; by the sheaf condition, this empty family can be glued into an element of A(o). Thus A(o) is non-empty and A(o) is a singleton. Finally this shows that the sheaf associated to a constant presheaf Q is the same as the sheaf associated to the presheaf Q' which coincides with Q at each $u \neq o$ and which is such that Q'(o) is a singleton.

For all these reasons and for the simplicity of the proof which follows, we now change our conventions slightly. $\mathbb{T}$ is the external integral theory. We associate with it a separated presheaf of theories which is $\mathbb{T}$ for each $u \neq o$ and the degenerate theory (a single operation in each dimension) when $u = o$. The corresponding category of algebraic presheaves is denoted by $\mathrm{Pr}(\mathbb{H}, \mathbb{T})$. Finally we consider the associated sheaf of theories and corresponding category $\mathrm{Sh}(\mathbb{H}, \mathbb{T})$ of algebraic sheaves. The previous remarks show that the sheaf of algebraic theories is exactly the sheaf associated to the constant presheaf on $\mathbb{T}$; in particular $\mathrm{Sh}(\mathbb{H}, \mathbb{T})$ is just the category defined previously (see I - 4).

Finally recall that we denote by $F : \mathrm{Pr}(\mathbb{H}) \to \mathrm{Pr}(\mathbb{H}, \mathbb{T})$ the left adjoint to the forgetful functor $\mathrm{Pr}(\mathbb{H}, \mathbb{T}) \to \mathrm{Pr}(\mathbb{H})$. Thus, if P is some presheaf on $\mathbb{H}$ and $v \in \mathbb{H}$

$$F\,P(v) = \begin{cases} F(Pv) & \text{if } v \neq o \\ \{*\} & \text{if } v = o \end{cases}$$

where $F(Pv)$ denotes the free $\mathbb{T}$-algebra on the set $P(v)$. In particular if $u \in \mathbb{H}$

$$F\,h_u(v) = \begin{cases} F1 & \text{if } o \neq v \leqslant u \\ F0 & \text{if } \quad v \not\leqslant u \\ \{*\} & \text{if } o = v. \end{cases}$$

This is a separated presheaf because $F\,h_u(o) = \{*\}$ and if $o \neq v = \bigvee_{i \in I} v_i$ in $\mathbb{H}$ and x, y are elements in $F\,h_u(v)$ such that for any i, $x|_{v_i} = y|_{v_i}$, then $x = y$. Indeed if $v \leqslant u$, choose $v_i \neq o$ and the restriction to v_i is just the identity on $F1$; now if $v \not\leqslant u$, choose $v_i \not\leqslant u$ and the restriction to v_i is just the identity on $F0$; in both cases, we deduce $x = y$. This separation property of $F\,h_u$ is the reason why we introduced this slightly different presentation.

Now consider a formal initial segment $\mathcal{U}$ of $\mathrm{Sh}(\mathbb{H}, \mathbb{T})$. We must prove that $\mathcal{U}$ is in fact the formal initial segment generated (via proposition II - 5) by some element $u \in \mathbb{H}$. By proposition II - 13, it suffices to prove the equality $\mathcal{U} = u$ in the larger frame $\mathrm{Heyt}(a\,F\,h_1)$, that is to say

$$\mathcal{U}_!\, \mathcal{U}^*(a\,F\,h_1) \cong a\,F\,h_u.$$

First, we need to construct u. $\mathbb{H}$ is a subframe of $\mathcal{H}$ and so we may define u by

$$u = \bigvee \{v \in \mathbb{H} \mid v \leqslant \mathcal{U}\}.$$

For any $v \in \mathbb{H}$, $v \leqslant u$, we have

$$\mathcal{U}_!\, \mathcal{U}^*(a\,F\,h_1)(v) \geqslant u_!\, u^*(a\,F\,h_1)(v) = a\,F\,h_u(v).$$

But $F\,h_1$ and $F\,h_u$ coincide for all $v \leqslant u$; thus $a\,F\,h_1$ and $a\,F\,h_u$ coincide for all $v \leqslant u$; in particular $a\,F\,h_u(v) = a\,F\,h_1(v)$ which proves finally that

$$\mathcal{U}_!\, \mathcal{U}^*(a\,F\,h_1)(v) = a\,F\,h_1(v) \qquad \text{if } v \leqslant u.$$

Now consider the following pullback in $\mathrm{Pr}(\mathbb{H}, \mathbb{T})$

$$\begin{array}{ccc} P & \rightarrowtail & \mathcal{U}_!\, \mathcal{U}^*(a\,F\,h_1) \\ \downarrow & & \downarrow \\ F\,h_1 & \rightarrowtail & a\,F\,h_1 ; \end{array}$$

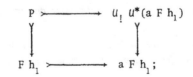

$F\,h_1$ is indeed a subobject of $a\,F\,h_1$ because $F\,h_1$ is a separated presheaf. Apply the associated sheaf functor to this diagram to obtain

$$\begin{array}{ccc} a\,P & \rightarrowtail & U_!\,U^*(a\,F\,h_1) \\ \downarrow & & \downarrow \\ a\,F\,h_1 & = & a\,F\,h_1; \end{array}$$

again this is a pullback which shows that $U_!\,U^*(a\,F\,h_1)$ is the sheaf associated to P. Combining this last result with the former, we obtain that for $v \leqslant u$ in H, $U_!\,U^*(a\,F\,h_1)(v) = a\,F\,h_1(v)$ and thus

$$P(v) = F\,h_1(v) = \begin{cases} F1 & \text{if} \quad o \neq v \leqslant u \\ \{*\} & \text{if} \quad o = v. \end{cases}$$

Comparing with the description of $F\,h_u$, we deduce a canonical factorization

$$\begin{array}{ccccc} & & P & & \\ & \nearrow & & \searrow & \\ F\,h_u & & \rightarrowtail & & F\,h_1 \end{array}$$

which, for $v \not\leqslant u$, is the unique morphism $F0 \to P(v)$. In particular, $F\,h_u$ is a subobject of P.

We will prove that $U_!\,U^*(a\,F\,h_1) = a\,P$ is just $a\,F\,h_u$. Suppose $a\,P \neq a\,F\,h_u$. Then P cannot be equal to $F\,h_u$ and thus there exists some $v \in H$ such that $F\,h_u(v)$ is not the whole $P(v)$. Now the descriptions of $P(v)$ and $F\,h_u(v)$ show that $v \not\leqslant u$. Then $F\,h_u(v) = F0$ and the inequality $F\,h_u(v) \lneq P(v)$ shows the existence of a 1-ary operation $\alpha \in P(v)$ which is not a constant. Denote by $\langle\alpha\rangle \subseteq F1$ the subalgebra of F1 generated by α. Define also the subpresheaf $\langle\langle\alpha\rangle\rangle$ of P generated by α, i.e.

$$\langle\langle\alpha\rangle\rangle\,(w) = \begin{cases} \langle\alpha\rangle & \text{if} \quad o \neq w \leqslant v \\ F0 & \text{if} \quad w \not\leqslant v \\ \{*\} & \text{if} \quad o = w. \end{cases}$$

By proposition I - 4, there is a morphism

$$f : F\,h_v \to \langle\langle\alpha\rangle\rangle$$

which corresponds to the choice of $\alpha \in F\,h_v(v)$. We can easily describe f :

(1) if $o \neq w \leqslant v$ $\quad f(w) : F1 \to \langle\alpha\rangle$ is the morphism which sends $*$ to α; by definition of α, this is a surjection. But the composite

$$F1 \xrightarrow{f(w)} \langle\alpha\rangle \rightarrowtail F1$$

is injective because α is non-constant (integrality of the theory; proposition 5); thus f(w) is injective. Finally f(w) is an isomorphism.

(2) if $w \not\leqslant v$ $\quad f(w) : F0 \to F0$ is the identity on F0; it is an isomorphism.

(3) if $w = o$ $\quad f(w) : \{*\} \to \{*\}$ is again the identity and thus an isomorphism. So, f is an isomorphism.

Finally, consider the following composite

$$a\ F\ h_v \underset{a(f^{-1})}{\cong} a \langle\langle\alpha\rangle\rangle \rightarrowtail a\ P = \mathcal{U}_!\ \mathcal{U}^*(a\ F\ h_1).$$

By proposition II - 13, this implies $v \leqslant \mathcal{U}$ in Heyt$(a\ F\ h_1)$ and thus in $\mathcal{H}$. By definition of u, this implies $v \leqslant u$, which contradicts $v \not\leqslant u$. Thus $\mathcal{U}_!\ \mathcal{U}^*(a\ F\ h_1) = a\ P = a\ F\ h_u$ and $\mathcal{U} = u$. Finally, $\mathcal{H} = H$ which proves the proposition. ■

Theorem 27 (Characterization theorem).

Consider an integral theory $\mathbb{T}$.

Two frames $\mathbf{H}$ *and* $\mathbf{H}'$ *are isomorphic*

if and only if

the categories Sh($\mathbf{H}, \mathbb{T}$) *and* Sh($\mathbf{H}', \mathbb{T}$) *are equivalent.*

By proposition 26. ■

CHAPTER 5 : SPECTRUM OF A THEORY

Recall that if R is a commutative ring, the Grothendieck - spectrum of R is some topological space X constructed from the prime ideals of R. The topological structure of X reflects some algebraic properties of R. Moreover, R can be presented as the ring of global sections of some sheaf of rings on X. The interest of this representation lies in the fact that the stálks of the sheaf are local rings. Thus the study of an arbitrary commutative ring can be reduced to the study of local rings if one accepts to replace a single ring by a sheaf of rings.

There are many other notions of spectrum and sheaf representation. For example, Pierce's representation is based on the properties of idempotents of R. It is simpler than Grothendieck's, but on the other hand Pierce needs assumptions on R to get interesting properties of the stalks : for regular rings, the stalks are fields.

In this chapter, we introduce, for an arbitrary algebraic theory $\mathbb{T}$, a notion of spectrum for $\mathbb{T}$ and a sheaf representation on this spectrum for any $\mathbb{T}$-algebra. This is obtained from the general theory of formal initial segments. In further chapters, we shall particularize these constructions to the case of a ring R, via the theory of R-modules.

If $\mathbb{H}$ is a frame and $\mathbb{T}$ is an algebraic theory in $\mathrm{Sh}(\mathbb{H})$, we saw in chapter 2 how to construct the frame $\mathcal{H}$ of formal initial segments of $\mathrm{Sh}(\mathbb{H}, \mathbb{T})$. This frame depends on $\mathbb{H}$ and on $\mathbb{T}$. But if we fix $\mathbb{H}$ to be the initial frame $\{0, 1\}$, thus if we work with sheaves on the one point topological space, the theory $\mathbb{T}$ is just a theory in the external sense and $\mathrm{Sh}(\mathbb{H}, \mathbb{T})$ is simply $\underline{\mathrm{Sets}}^{\mathbb{T}}$. In other words, to any algebraic theory $\mathbb{T}$, we can associate a frame $\mathcal{H}_{\mathbb{T}}$ which depends only on $\mathbb{T}$: this is the frame of formal initial segments of $\underline{\mathrm{Sets}}^{\mathbb{T}}$. In this chapter, we prove that this frame is spatial, i.e. is the frame of open subsets of some topological space called the spectrum of $\mathbb{T}$.

The inclusion of frames $\mathbb{H} \hookrightarrow \mathcal{H}$ studied in chapter 2 reduces here to $\{0, 1\} \hookrightarrow \mathcal{H}_{\mathbb{T}}$ and the restriction functor

$$\Gamma : \mathrm{Sh}(\mathcal{H}_{\mathbb{T}}, \mathbb{T}) \to \mathrm{Sh}(\{0, 1\}, \mathbb{T}) \cong \underline{\mathrm{Sets}}^{\mathbb{T}}$$

is simply the "global sections" functor. In proposition II - 14, we produced a right inverse Δ to this functor. In other words, for any $\mathbb{T}$-algebra A, there is a sheaf ΔA of $\mathbb{T}$-algebras on the spectrum of $\mathbb{T}$, whose algebra of global sections is A. This is a sheaf representation theorem for $\mathbb{T}$-algebras.

Considering the results of chapter 4, we deduce that for an integral theory $\mathbb{T}$, the spectrum of $\mathbb{T}$ is just the one point set and therefore the corresponding representation theory vanishes in that case. Thus the theory developed in this chapter will be useful in the case of non-integral theories.

§ 1. THE PURE SPECTRUM OF AN ALGEBRAIC THEORY

In this paragraph, we prove that the frame of formal initial segments of $\text{Sets}^{\mathbb{T}}$, where $\mathbb{T}$ is any theory, is the frame of open subsets of a compact space (not necessarily Hausdorff).

Definition 1.

If $\mathbb{T}$ *is any algebraic theory,* $\mathcal{K}_{\mathbb{T}}$ *denotes the frame of formal initial segments of* $\text{Sets}^{\mathbb{T}}$.

We still denote by FX the free $\mathbb{T}$-algebra on the set X.

Definition 2.

If $\mathbb{T}$ *is any algebraic theory, a* $\mathbb{T}$*-ideal is a sub-* $\mathbb{T}$*-algebra of* F1.

If R is a ring with unit and $\mathbb{T}$ is the theory of left R-modules, F1 is simply R and a left-submodule of R is simply a left ideal of R. This justifies our terminology.

Definition 3.

If $\mathbb{T}$ *is any algebraic theory, a* $\mathbb{T}$*-ideal is pure if it is of the form* $U_! U^*(F1)$ *for some* U *in* $\mathcal{K}_{\mathbb{T}}$.

We denote by $p(\mathbb{T})$ the set of pure $\mathbb{T}$-ideals. In chapter 6, we will see that in the case of left modules on a ring R with unit, definition 3 reduces to the usual definition of a pure ideal.

Proposition 4.

Any pure $\mathbb{T}$*-ideal is a Heyting subobject of* F1.

By definition of a formal initial segment (II - 6). ■

Proposition 5.

The pure $\mathbb{T}$*-ideals constitute, for the usual operations of union and intersection of sub-*$\mathbb{T}$*-algebras, a frame. Moreover,* $p(\mathbb{T})$ *is isomorphic to* $\mathcal{H}_{\mathbb{T}}$*.*

By proposition II - 12, the inclusion $\mathcal{H}_{\mathbb{T}} \hookrightarrow \mathrm{Heyt}(F1)$ is the map sending the formal initial segment U to the pure $\mathbb{T}$-ideal $U_! \, U^*(F1)$. ■

Definition 6.

A pure $\mathbb{T}$*-ideal is purely maximal if it is maximal among the proper pure* $\mathbb{T}$*-ideals.*

Definition 7.

A pure $\mathbb{T}$*-ideal* J *is purely prime if it is proper and for any pure* $\mathbb{T}$*-ideals* I_1, I_2

$$I_1 \cap I_2 \subseteq J \quad \Rightarrow \quad I_1 \subseteq J \text{ or } I_2 \subseteq J.$$

Proposition 8.

Any purely maximal $\mathbb{T}$*-ideal is purely prime.*

Suppose $J \subseteq F1$ is a purely maximal $\mathbb{T}$-ideal and I_1, I_2 are pure $\mathbb{T}$-ideals such that $I_1 \cap I_2 \subseteq J$. If $I_1 \not\subseteq J$, then $I_1 \cup J$ (union as sub-$\mathbb{T}$-algebras of F1) is still a pure ideal (proposition 5) and is strictly larger than J; the maximality of J implies $I_1 \cup J = F1$. But then, by proposition 4

$$\begin{aligned} I_2 &= I_2 \cap F_1 \\ &= I_2 \cap (I_1 \cup J) \\ &= (I_2 \cap I_1) \cup (I_2 \cap J) \\ &\subseteq J \cup J \\ &= J \end{aligned}$$

which concludes the proof. ■

Proposition 9.

Any proper pure $\mathbb{T}$*-ideal is contained in a purely maximal* $\mathbb{T}$*-ideal.*

Let $I \subseteq F1$ be a pure $\mathbb{T}$-ideal, $I \neq F1$. Consider the set X, ordered by inclusion

$$X = \{J \mid J \text{ pure } \mathbb{T}\text{-ideal; } I \subseteq J \neq F1\}$$

X is not empty because I is some element in X. If $(J_k)_{k \in K}$ is a totally ordered family of elements in X, $\bigcup_{k \in K} J_k$ is again a pure ideal containing I. Now for any k, $1 \notin J_k$ and the union $\bigcup_{k \in K} J_k$ is filtered : it is thus exactly the union of the underlying sets ([21] - 18 - 8 - 8) and $1 \notin \bigcup_{k \in K} J_k$. So we can apply Zorn's lemma to X and obtain a proper $\mathbb{T}$-ideal J which is maximal among the proper pure ideals containing I. But obviously this implies that J is purely maximal. ■

Proposition 10.

If I *is a pure* $\mathbb{T}$*-ideal and* $a \in F1 \setminus I$, *there exists a purely prime* $\mathbb{T}$*-ideal* J *such that* $I \subseteq J$ *and* $a \notin J$.

Consider the set, ordered by inclusion,

$$X = \{J \mid J \text{ pure } \mathbb{T}\text{-ideal, } I \subseteq J,\ a \notin J\}.$$

X is not empty because $I \in X$. Now if $(J_k)_{k \in K}$ is a totally ordered family of elements in X, just as in the proof of proposition 9, $\bigcup_{k \in K} J_k$ is a pure $\mathbb{T}$-ideal such that $a \notin \bigcup_{k \in K} J_k$. We may then apply Zorn's lemma and choose some J maximal in X.

We assert that J is purely prime. By contraposition, take I_1, I_2 pure with $I_1 \not\subseteq J$ and $I_2 \not\subseteq J$; we must show that $I_1 \cap I_2 \not\subseteq J$. From $I_k \not\subseteq J$, we deduce that $I_k \cup J$ is a pure $\mathbb{T}$-ideal containing strictly J; by definition of X, this implies $a \in I_k + J$. Finally $a \in (I_1 \cap I_2) + J$ and $a \notin J$, thus $I_1 \cap I_2 \not\subseteq J$. ■

Proposition 11.

Any proper pure $\mathbb{T}$*-ideal is the intersection of the purely prime* $\mathbb{T}$*-ideals containing it.*

By propositions 8 - 9, there are purely prime ideals containing a proper pure ideal I and by proposition 10, any element which is not in I is outside one of them, thus outside their intersection. This proves the proposition. ■

We turn now to the construction of the pure spectrum of $\mathbb{T}$. We consider first the set $pp(\mathbb{T})$ of purely prime $\mathbb{T}$-ideals and its power set $2^{pp(\mathbb{T})}$. We shall define a topology on $pp(\mathbb{T})$ and the corresponding topological space

$Spp(\mathbb{T})$ will be the pure spectrum of $\mathbb{T}$. In order to do so, we define a mapping :

$$\mathcal{O} : p(\mathbb{T}) \longrightarrow 2^{pp(\mathbb{T})}$$

by $\mathcal{O}(I) = \{J \in pp(\mathbb{T}) \mid I \not\subseteq J\}$.

Proposition 12.

$\mathcal{O}$ is a morphism of frames.

(1) smallest element :

$$\mathcal{O}(F0) = \{J \in pp(\mathbb{T}) \mid F0 \not\subseteq J\} = \phi.$$

(2) greatest element :

$$\mathcal{O}(F1) = \{J \in pp(\mathbb{T}) \mid F1 \not\subseteq J\} = pp(\mathbb{T}).$$

(3) finite intersections :

Consider I_1, I_2 in $p(\mathbb{T})$ and J in $Spp(\mathbb{T})$. Since J is purely prime, we have the equivalence

$$I_1 \cap I_2 \not\subseteq J \Leftrightarrow I_1 \not\subseteq J \text{ and } I_2 \not\subseteq J.$$

Therefore

$$\begin{aligned} \mathcal{O}(I_1 \cap I_2) &= \{J \in pp(\mathbb{T}) \mid I_1 \cap I_2 \not\subseteq J\} \\ &= \{J \in pp(\mathbb{T}) \mid I_1 \not\subseteq J \text{ and } I_2 \not\subseteq J\} \\ &= \{J \in pp(\mathbb{T}) \mid I_1 \not\subseteq J\} \cap \{J \in pp(\mathbb{T}) \mid I_2 \not\subseteq J\} \\ &= \mathcal{O}(I_1) \cap \mathcal{O}(I_2). \end{aligned}$$

(4) arbitrary unions :

Let $(I_k)_{k\in K}$ be a family of elements in $p(\mathbb{T})$. Then

$$\begin{aligned} \mathcal{O}(\bigcup_{k\in K} I_k) &= \{J \in pp(\mathbb{T}) \mid \bigcup_{k\in K} I_k \not\subseteq J\} \\ &= \{J \in pp(\mathbb{T}) \mid \exists k \in K \; I_k \not\subseteq J\} \\ &= \bigcup_{k\in K} \{J \in pp(\mathbb{T}) \mid I_k \not\subseteq J\} \\ &= \bigcup_{k\in K} \mathcal{O}(I_k). \end{aligned}$$

■

Proposition 13.

$\mathcal{O}$ is injective.

Consider $I_1 \neq I_2$ in $p(\mathbb{T})$. One of these ideals is not contained in the other one; suppose $I_1 \not\subseteq I_2$ and choose $a \in I_1 \setminus I_2$. By proposition 10, we can choose a purely prime $\mathbb{T}$-ideal J such that $I_2 \subseteq J$ and $a \notin J$; in particular $I_1 \not\subseteq J$. But this implies $J \in \mathcal{O}(I_1)$ and $J \notin \mathcal{O}(I_2)$. Thus $\mathcal{O}(I_1) \neq \mathcal{O}(I_2)$ and $\mathcal{O}$ is injective.

■

Proposition 14.

The subsets $O(\mathrm{I})$, *for* I *running through* $p(\mathbb{T})$, *constitute a topology on* $pp(\mathbb{T})$.

By proposition 12, this family of subsets contains ϕ, F1 and is stable for finite intersections and arbitrary unions. ■

Definition 15.

The pure spectrum $Spp(\mathbb{T})$ *of the theory* $\mathbb{T}$ *is the set* $pp(\mathbb{T})$ *of purely prime* $\mathbb{T}$*-ideals equipped with the topology whose open subsets are the* $O(\mathrm{I})$*'s, for any pure* $\mathbb{T}$*-ideal* I.

Proposition 16.

The frame of open subsets of the pure spectrum $Spp(\mathbb{T})$ *of* $\mathbb{T}$ *is isomorphic to the frame* $p(\mathbb{T})$ *of pure* $\mathbb{T}$*-ideals.*

By propositions 12 - 13. ■

Theorem 17.

The pure spectrum $Spp(\mathbb{T})$ *of a theory* $\mathbb{T}$ *is a compact space.*

Consider a family $(I_k)_{k\in I}$ of pure $\mathbb{T}$-ideals such that $\bigcup_{k\in K} O(I_k) = Spp(\mathbb{T})$. By proposition 16, this is equivalent to $\bigcup_{k\in K} I_k = F1$. But $\bigcup_{k\in K} I_k$ (union as $\mathbb{T}$-algebras) is the sub-$\mathbb{T}$-algebra of F1 generated by all the elements of all the I_k's. An element in $\bigcup_{k\in I} I_k$ has thus the form $\alpha(a_{k_1}, \ldots, a_{k_n})$ where $a_{k_i} \in I_{k_i}$, n is any integer and α is a n-ary operation. The equality $\bigcup_{k\in K} I_k = F1$ implies that the universal generator $*$ of F1 is of the form $\alpha(a_{k_1}, \ldots, a_{k_n})$. But this implies that $*$ is in $\bigcup_{i=1}^{n} I_{k_i}$ and thus also any $\beta(*) = \beta$ for any 1-ary operation $\beta \in F1$. Finally, $F1 = \bigcup_{i=1}^{n} I_{k_i}$ and $Spp(\mathbb{T}) = O(F1) = \bigcup_{i=1}^{n} O(I_{k_i})$. This concludes the proof. ■

The spectrum $Spp(\mathbb{T})$ is generally not a Hausdorff space. A counterexample will be given in chapter 7 when $\mathbb{T}$ is the theory of right modules on the ring R of linear endomorphisms of a vector space with infinite countable dimension.

§ 2. REPRESENTATION THEOREM FOR $\mathbb{T}$-ALGEBRAS

In this paragraph, we interpret some results of chapter 2 to present any $\mathbb{T}$-algebra as the algebra of global sections of a sheaf of $\mathbb{T}$-algebras on the pure spectrum of $\mathbb{T}$.

Proposition 18.

The "global sections" functor

$$\Gamma : \mathrm{Sh}(\mathrm{Spp}(\mathbb{T}), \mathbb{T}) \to \underline{\mathrm{Sets}}^{\mathbb{T}}$$

has a right inverse Δ.

$\mathrm{Sh}(\mathrm{Spp}(\mathbb{T}), \mathbb{T})$ is simply $\mathrm{Sh}(\mathcal{K}_{\mathbb{T}}, \mathbb{T})$ and $\underline{\mathrm{Sets}}^{\mathbb{T}}$ is simply $\mathrm{Sh}(\{0, 1\}, \mathbb{T})$. Moreover, $\mathcal{K}_{\mathbb{T}}$ is the frame of formal initial segments of $\mathrm{Sh}(\{0, 1\}, \mathbb{T})$. The "global sections" functor

$$\Gamma : \mathrm{Sh}(\mathcal{K}_{\mathbb{T}}, \mathbb{T}) \to \mathrm{Sh}(\{0, 1\}, \mathbb{T})$$

is given by

$\Gamma(A)(0) = \{*\}$

$\Gamma(A)(1) = A(1)$.

Thus $\Gamma(A)$ is the restriction of A to $\{0, 1\}$. Therefore the result follows immediately from proposition II - 14. ■

Theorem 19.

For any $\mathbb{T}$*-algebra* A, A *is isomorphic to the* $\mathbb{T}$*-algebra of global sections of the sheaf* ΔA *on the pure spectrum* $\mathrm{Spp}(\mathbb{T})$ *of* $\mathbb{T}$.

This is another way of saying that Δ is right inverse to Γ. ■

CHAPTER 6 : APPLICATIONS TO MODULE THEORY

In this chapter, we apply our previous results to the particular case of module theory on a ring R. Thus $\mathbb{T}$ will be the theory of right R-modules, where R is a ring with unit. The frame $\mathbb{H}$ will again be the initial frame $\{0, 1\}$ so that all the sheaves we consider are sheaves on the singleton. ($\{0, 1\}$ is the frame of open subsets of the singleton). Thus the category $Sh(\mathbb{H}, \mathbb{T})$ is simply the category $\underline{Mod}_R$ of right R-modules.

If R is commutative, the theory of R-modules is a commutative theory and the results of chapter 3 can be applied. They imply the existence of a topos $\mathcal{E}(\mathbb{H}, \mathbb{T})$, which is the topos of $(R, \times)$-sets, in which lives an object $\Omega_{\mathbb{T}}$. This object is the set of right ideals of R. Each submodule $B \rightarrowtail A$ can be classified by a morphism $\varphi : M \to \Omega_{\mathbb{T}}$ and each localization of $\underline{Mod}_R$ can be classified by a "topology" $j : \Omega_{\mathbb{T}} \to \Omega_{\mathbb{T}}$.

Now if R is any ring with unit, the results of chapter 5 apply to the theory of right R-modules. This produces a "pure spectrum of R" and for any R-module A, a sheaf representation of A on the pure spectrum of R. This spectrum and this representation are thus obtained via the formal open subsets of $\underline{Mod}_R$, which are characterized by certain ideals of R; in chapter 5, we called them "pure $\mathbb{T}$-ideals". Here we show that in the case of R-modules, they are exactly the pure ideals of the ring in the usual sense of module theory.

R itself is a R-module. Thus it can be represented by a sheaf of R-modules on its pure spectrum. In fact this sheaf of R-modules is also a sheaf of rings. Moreover, the sheaf representation of any R-module is a sheaf of modules on the sheaf of rings which represents R.

§ 1. THE CLASSIFYING OBJECT FOR MODULE THEORY

Consider a commutative ring with unit R and the corresponding theory $\mathbb{T}$ of modules. Consider the frame $\mathbb{H} = \{0, 1\}$ of open subsets of the singleton. Then $Sh(\mathbb{H}, \mathbb{T})$ is the category $\underline{Mod}_R$ of R-modules. The monoid of 1-ary operations of $\mathbb{T}$ is the multiplicative monoid $(R, \times)$. Thus the topos $\mathcal{E}(\mathbb{H}, \mathbb{T})$ described in chapter 3 is the topos $\underline{Sets}_R$ of $(R, \times)$-sets.

The two representable algebras are a F h_0 = (0) and a F h_1 = R. Thus $\Omega_{\mathbb{T}}$ is the set Ω_R of submodules of R, i.e. the set of ideals of R. Ω_R has the structure of a (R, ×)-set via

$$\Omega_R \times R \longrightarrow R$$
$$(I, r) \longmapsto r^{-1}(I) = [I : r] .$$

Ω_R is ordered by inclusion of ideals, has greatest element R and finite intersections.

Now if $B \rightarrowtail A$ is the inclusion of some submodule, the results of chapter 3 tell us also how to construct the corresponding characteristic map $\varphi : A \to \Omega_{\mathbb{T}}$. If $a \in A$, consider the corresponding linear mapping $\bar{a} : R \to A$ such that $\bar{a}(1) = a$. Then $\varphi(a)$ is $\bar{a}^{-1}(B)$, i.e.

$$\varphi(a) = \{r \in R \mid a\, r \in B\}.$$

In other words, the ideals of R are the truth-values of the theory of R-modules and for $a \in A$, $B \subseteq A$, the truth value of $a \in B$ is the ideal constituted by those r such that $a\, r \in B$ is true.

By theorem III - 56, there is a one-to-one correspondance between localizations of $\underline{\mathrm{Mod}}_R$ and $\mathbb{T}$-topologies on $\Omega_{\mathbb{T}}$. A $\mathbb{T}$-topology on $\Omega_{\mathbb{T}}$ is a mapping $j : \Omega_R \to \Omega_R$ in $\underline{\mathrm{Sets}}_R$, satisfying three conditions. More precisely, a $\mathbb{T}$-topology on Ω_R is a mapping

$$j : \Omega_R \longrightarrow \Omega_R$$

such that for $I, J \in \Omega_R$ and $r \in R$

(1) $j([I : r]) = [j(I) : r]$

(2) $j(R) = R$

(3) $j\, j(I) = j(I)$

(4) $j(I \cap J) = j(I) \cap j(J)$.

This is exactly what H. Simmons calls a "localizor" (cfr. [22]).

If P is a prime ideal in R and $R \to R_P$ is the usual localization of R at P, this morphism of rings induces a full and faithful functor $\pi : \underline{\mathrm{Mod}}_{R_P} \to \underline{\mathrm{Mod}}_R$ between the corresponding categories of modules. But R_P is a flat R-module; thus the functor $- \otimes_R R_P : \underline{\mathrm{Mod}}_R \to \underline{\mathrm{Mod}}_{R_P}$ is exact. On the other hand, $- \otimes_R R_P$ is left adjoint to π, so that $\underline{\mathrm{Mod}}_{R_P}$ is a localization of $\underline{\mathrm{Mod}}_R$.

This localization is thus classified by a $\mathbb{T}$-topology $j_P : \Omega_R \to \Omega_R$. We now proceed to describe this topology.

The localization I_P of an ideal I is the set of fractions

$$I_P = \{\frac{i}{q} \mid i \in I;\ q \notin P\}$$

where $\frac{i}{q} = \frac{i'}{q'}$ if $iq' = i'q$. Therefore, if $I \subseteq J$ are two ideals, $I_P = J_P$ if and only if any $\frac{j}{q}$ with $j \in J$, $q \notin P$ can be written as a fraction $\frac{i}{q'}$ with $i \in I$, $q' \notin P$. This is equivalent to the existence of an element $r \notin P$ such that $j\,r \in I$. Indeed, if such an r exists, $\frac{j}{q} = \frac{j\,r}{q\,r}$ with $j\,r \in I$ and $q\,r \notin P$, since P is prime. Conversely if $\frac{j}{q} = \frac{i}{q'}$, $j\,q' = i\,q \in I$ and $\frac{j\,q'}{q\,q'} = \frac{j}{q}$ with $j\,q' \in I$ and $q\,q' \notin P$, since P is prime. Thus

$$I_P = J_P \iff \forall j \in J \quad \exists r \notin P \quad j\,r \in I.$$

But the closure $\bar{I}$ of I for the universal closure operation associated to j_P is $j_P(I)$ and this is the largest ideal J such that $I_P = J_P$. Therefore

$$j_P(I) = \{j \in R \mid \exists r \notin P \quad j\,r \in I\}.$$

Indeed, the only thing we still need to prove is that $j_P(I)$ is an ideal. If $j\,r \in I$ and $j'\,r' \in I$ with $r, r' \notin P$, then $j\,r\,r' \in I$ and $j'\,r'\,r \in I$ with $r\,r' \notin P$ (P is prime) and $(j + j')(r\,r') \in I$. So $j_P(I)$ is stable under addition. It is obviously stable under multiplication by some element of R. So it is an ideal.

§ 2. THE PURE IDEAL ASSOCIATED TO A FORMAL INITIAL SEGMENT

Consider any ring with unit R (not necessarily commutative) and the category $\underline{Mod}_R$ of right R-modules. By proposition II - 12, any formal initial segment of $\underline{Mod}_R$ can be completely characterized by some Heyting subobject of R in $\underline{Mod}_R$, i.e. by some (special) right ideal of R. If U is the formal initial segment, this ideal I is $U_! \, U^* \, R$. In this paragraph we will prove that I is a 2-sided left-pure ideal of R.

Definition 1 (cfr. [5]).

A submodule $B \rightarrowtail A$ *in the category of left* R*-modules is called right pure if, for any right* R*-module* M *the canonical morphism*

$$M \otimes B \longrightarrow M \otimes A$$

is injective.

Before stating the next proposition, observe that any morphism $0 \to A$ is injective in $\underline{\mathrm{Mod}}_R$ (cfr. definition II - 6).

Proposition 2.

If $\mathcal{U}$ is a formal initial segment of $\underline{\mathrm{Mod}}_R$ characterized by the ideal $I = \mathcal{U}_! \, \mathcal{U}^* R$, the objects of the full subcategory $\mathcal{U}$ are those modules A such that the canonical morphism $A\,I \to I$ is an isomorphism; moreover $\mathcal{U}^*(A) = A\,I$.

All modules and ideals are defined on the right. $\mathcal{U}_! : \mathcal{U} \hookrightarrow \underline{\mathrm{Mod}}_R$ is the canonical inclusion. $\mathcal{U}^* : \underline{\mathrm{Mod}}_R \to \mathcal{U}$ is its right adjoint and $\mathcal{U}_* : \mathcal{U} \to \underline{\mathrm{Mod}}_R$ is right adjoint to $\mathcal{U}^*$. $\mathcal{U}_!$ and $\mathcal{U}^*$ are thus cocontinuous and $\mathcal{U}_! \, \mathcal{U}^* R = I$. We know that $I\,R = I$, since I is a right ideal.

Consider a free module $F(K)$ on a set K of generators.

$$\begin{aligned}
\mathcal{U}_! \, \mathcal{U}^* F(K) &= \mathcal{U}_! \, \mathcal{U}^* F(\coprod_K 1). \\
&= \mathcal{U}_! \, \mathcal{U}^*(\coprod_K F1) \\
&= \coprod_K \mathcal{U}_! \, \mathcal{U}^* F1 \\
&= \coprod_K \mathcal{U}_! \, \mathcal{U}^* R \\
&= \coprod_K I \\
&= \left\{ (i_k)_{k \in K} \;\middle|\; \begin{array}{l} i_k \in I \text{ is zero except for a finite} \\ \text{number of indexes} \end{array} \right\}
\end{aligned}$$

Now any module A is a quotient of the free module FA. We then have the following commutative diagram

$$\begin{array}{ccc}
\mathcal{U}_! \, \mathcal{U}^*(FA) & \rightarrowtail & FA \\
{\scriptstyle \mathcal{U}_! \, \mathcal{U}^*(\Sigma)} \downarrow & & \downarrow {\scriptstyle \Sigma} \\
\mathcal{U}_! \, \mathcal{U}^*(A) & \rightarrowtail & A,
\end{array}$$

and $\mathcal{U}_! \, \mathcal{U}^*(A)$ is the image of

$$\mathcal{U}_! \, \mathcal{U}^*(FA) \rightarrowtail FA \xrightarrow{\Sigma} A$$

where Σ is defined by :

$$FA = \left\{ (r_a)_{a\in A} \;\middle|\; \begin{array}{l} r_a \in R \text{ is zero except for a finite number} \\ \text{of indexes} \end{array} \right\}$$

$$\Sigma((r_a)_{a\in A}) = \sum_{a\in A} a \,.\, r_a .$$

The description of $\mathcal{U}_!\, \mathcal{U}^*(FA)$ implies that

$$\mathcal{U}_!\, \mathcal{U}^* A = \left\{ \sum_{a\in A} a \,.\, i_a \;\middle|\; \begin{array}{l} i_a \in I \text{ is zero except for a finite number} \\ \text{of indexes} \end{array} \right\}$$

$$= A \,.\, I.$$

So $\mathcal{U}^* A = A \,.\, I$. Now, a module A is in $\mathcal{U}$ if and only if $\mathcal{U}_!\, \mathcal{U}^* A = A$ (cfr. [21]), i.e. if and only if $A \,.\, I = A$. This concludes the proof. ■

Corollary 3.

Let I *characterize the formal initial segment* $\mathcal{U}$ *of* $\underline{\text{Mod}}_R$. *Then* I *is 2-sided.*

$I = \mathcal{U}_!\, \mathcal{U}^* R = R\, I$ (proposition 2). ■

Proposition 4.

Let I *characterize the formal initial segment* $\mathcal{U}$ *of* $\underline{\text{Mod}}_R$. *Then for every finite family* $(i_k)_{1\leqslant k\leqslant n}$ *of elements in* I, *there exists* $\varepsilon \in I$ *such that for all* k *we have* $i_k\, \varepsilon = i_k$.

Consider $(i_1, \ldots, i_n)R$ the submodule of $\coprod_{k=1}^{n} I$ generated by $(i_1, \ldots, i_n)$. Then

$$\begin{aligned}(i_1, \ldots, i_n)R \,.\, I &= (i_1, \ldots, i_n)R \cap \mathcal{U}_!\, \mathcal{U}^*(\coprod_{k=1}^{n} I) \\ &= (i_1, \ldots, i_n)R \cap \coprod_{k=1}^{n} (\mathcal{U}_!\, \mathcal{U}^* I) \\ &= (i_1, \ldots, i_n)R \cap (\coprod_{k=1}^{n} I) \\ &= (i_1, \ldots, i_n)R.\end{aligned}$$

Therefore we can write

$$\begin{aligned}(i_1, \ldots, i_n) &= \sum_{\ell} (i_1, \ldots, i_n) r_\ell\, j_\ell \\ &= (i_1, \ldots, i_n) \,.\, (\sum_{\ell} r_\ell\, j_\ell)\end{aligned}$$

where $r_\ell \in R$ and $j_\ell \in I$; thus $\varepsilon = \sum_\ell r_\ell j_\ell \in I$ by corollary 3. ∎

Proposition 5.

Let I characterize the formal initial segment $\mathcal{U}$ of $\underline{Mod}_R$. Then I, as a left module of R, is right pure.

Consider any $A \in \underline{Mod}_R$ and the canonical morphism $A \otimes I \to A \otimes R \cong A$ which sends $a \otimes i$ to $a\, i$. We must prove it is injective (definition 1). Take $\sum_k a_k \otimes i_k \in A \otimes I$ such that $\sum_k a_k\, i_k = 0$. By proposition 4, choose ε in I such that for any k, $i_k \cdot \varepsilon = i_k$. This implies that

$$\begin{aligned}\sum_k a_k \otimes i_k &= \sum_k (a_k \otimes i_k \cdot \varepsilon)\\ &= \sum_k (a_k \cdot i_k \otimes \varepsilon)\\ &= (\sum_k a_k \cdot i_k) \otimes \varepsilon\\ &= 0 \otimes \varepsilon\\ &= 0.\end{aligned}$$

This concludes the proof. ∎

§ 3. FORMAL INITIAL SEGMENT ASSOCIATED TO A PURE IDEAL

Again we work in the category $\underline{Mod}_R$ of right R-modules on the ring with unit R. We fix a 2-sided ideal I of R which is right pure when we regard it as a left submodule of R. We will construct a formal initial segment $\mathcal{U}$ of $\underline{Mod}_R$, starting from I. Since I is 2-sided, $A \otimes I$ is a right module for any $A \in \underline{Mod}_R$ (cfr. [20]).

Proposition 6.

Let I be a 2-sided ideal of R.

I is right pure if and only if, for any $A \in \underline{Mod}_R$, the canonical morphism $A \otimes I \to A\, I$ is an isomorphism.

The morphism $A \otimes I \to A \otimes R = R$ has image $A\, I$; thus it is injective if and only if $A \otimes I \to A\, I$ is bijective. ∎

Proposition 7.

Let I *be a 2-sided right pure ideal of* R. *Then for every finite family* $(i_k)_{1 \leq k \leq n}$ *of elements in* I, *there exists* $\varepsilon \in I$ *such that for all* k *we have* $i_k \varepsilon = i_k$.

Consider the cokernel

$$(i_1, \ldots, i_n)R \longrightarrow R^n \longrightarrow\!\!\!\!\to \frac{R^n}{(i_1, \ldots, i_n)R} \longrightarrow 0.$$

Tensoring on the right by I preserves cokernels; by proposition 6, we then obtain

$$(i_1, \ldots, i_n)R \,.\, I \longrightarrow R^n \,.\, I \longrightarrow\!\!\!\!\to \frac{R^n}{(i_1, \ldots, i_n).R} \,.\, I \to 0$$

and since I is 2-sided :

$$(i_1, \ldots, i_n)I \longrightarrow I^n \longrightarrow\!\!\!\!\to \frac{I^n}{(i_1, \ldots, i_n)R} \longrightarrow 0.$$

But $(i_1, \ldots, i_n) \in I^n$ is sended to 0 in the quotient; thus it comes from some element in $(i_1, \ldots, i_n)I$:

$$(i_1, \ldots, i_n) = (i_1, \ldots, i_n).\,\varepsilon \text{ with } \varepsilon \in I.$$ ■

Proposition 8.

Let I *be a 2-sided right pure ideal of* R. *For any* $A \in \underline{\mathrm{Mod}}_R$ *we have*

$$A\,I = \{a\,i \mid a \in A;\ i \in I;\ a\,i = a\}.$$

An element in A I has the form $\sum_k a_k i_k$ with $a_k \in A$ and $i_k \in I$. By proposition 7 choose $\varepsilon \in I$ such that for any k, $i_k \varepsilon = i_k$.

$$\sum_k a_k i_k = \sum_k (a_k i_k \varepsilon) = (\sum_k a_k i_k)\varepsilon = a\,\varepsilon,$$

with $\varepsilon \in I$. ■

Proposition 9.

Let I *be a 2-sided right pure ideal of* R. *The full subcategory* $\mathcal{U}$ *of* $\underline{\mathrm{Mod}}_R$ *whose objects are those modules* M *such that* M = M I *is a formal initial segment* $\mathcal{U}$ *of* $\underline{\mathrm{Mod}}_R$. *Moreover* $I = \mathcal{U}_! \, \mathcal{U}^* R$.

We denote the inclusion by $\mathcal{U}_! : \mathcal{U} \hookrightarrow \underline{\mathrm{Mod}}_R$. We define $\mathcal{U}^* : \underline{\mathrm{Mod}}_R \to \mathcal{U}$ by $\mathcal{U}^* = - \otimes I$. This makes sense since I . I = I by proposition 7 and thus for $A \in \underline{\mathrm{Mod}}_R$

$$(A \otimes I) . I = A . I . I = A . I = A \otimes I$$

by proposition 6. On the other hand, $A \otimes I$ is a right module since I is a right module. The adjunction $\mathcal{U}_! \dashv \mathcal{U}^*$ holds : indeed, consider $A \in \underline{Mod}_R$ and $B \in \mathcal{U}$; we must prove there is a bijection

$$(B, A) \cong (B, A . I).$$

Any morphism $B \to A.I$ is in particular a morphism $B \to A . I \subseteq A$. On the other hand, if $f : B \to A$ and $b \in B$, then $b \in B = B . I$ and thus $b = \sum_k b_k i_k$ with $i_k \in I$; so $f(b) = \sum_k f(b_k) i_k \in A . I$ and f factors through $A . I$. Thus $\mathcal{U}^*$ is right adjoint to $\mathcal{U}_!$. But $\mathcal{U}^* = - \otimes I$ has itself a right adjoint $\mathcal{U}_* = (I, -)$ because I is 2-sided (cfr. [20]).

We still need to prove condition (F 3 - 4 - 5) of definition II - 6. (F 3) means that if B is in $\mathcal{U}$, any submodule A of B is in $\mathcal{U}$. Indeed take $a \in A$; a is in $B = BI$ thus $a = ai$ with $i \in I$ (proposition 8). So a is in AI.

(F4) is obvious : the canonical morphism $A I \to A$ is simply the inclusion, for any $A \in \underline{Mod}_R$. It is also a Heyting subobject (F4). Indeed, (H3) is satisfied since $\underline{Mod}_R$ is an abelian category (cfr. [21] - 14 - 2 - 7). Now if S and T are any submodules of A, the inclusions

$$(A I \cap S) \cup (A I \cap T) \subseteq A I \cap (S \cup T)$$
$$(S \cap T) \cup (S \cap A I) \subseteq S \cap (T \cup A I)$$

certainly hold because they do for any subobjects. Now take $a \in A I \cap (S \cup T)$. From $a \in A I$, we deduce $a = a i$ with $i \in I$ (proposition 8) and from $a \in S \cup T$, we deduce $a = s + t$ with $s \in S$ and $t \in T$. Thus

$$a = a i = (s + t)i = s i + t i.$$

But $s i \in S$ because S is right sided and $s i \in I$ because I is left sided; so $s i \in S \cap I$ and in the same way $t i \in T \cap I$. Finally a is in $(S \cap I) \cup (T \cap I)$. This proves (H3). We verify (H4) in an analogous way : choose s in $S \cap (T \cup AI)$. From $s \in T \cup A I$ and proposition 8, we deduce

$$s = t + a i; \; t \in T; \; a \in A; \; i \in I.$$

From proposition 7 choose $\varepsilon \in I$ with $i \varepsilon = i$. So

$$\begin{aligned} s &= t + a i \\ &= (t - t \varepsilon) + (t + a i)\varepsilon \\ &= t(1 - \varepsilon) + s \varepsilon. \end{aligned}$$

Now we have $s \varepsilon \in S I \subseteq S \cap A I$. On the other hand, $t(1 - \varepsilon) \in T$ since $t \in T$ and $t(1 - \varepsilon) = s - s \varepsilon \in S$ since $s \in S$; so $t(1 - \varepsilon) \in S \cap T$. Therefore, s is in

$(S \cap T) \cup (S \cap A\, I)$. This proves (H 4).

Finally $u_! \, u^* \, I = I \otimes I = I \, . \, I = I$ by propositions 6 - 7. ■

§ 4. PURE SPECTRA OF A RING

For the theory of R-modules, the results of §§ 2 - 3 imply that a pure ideal in the sense of definition V - 3 is simply a 2-sided right-pure ideal of R. This yields an easier description of the pure spectrum of the theory of right R-modules : we call it simply the right pure spectrum of R.

Proposition 10.

If R *is a ring with unit and* $\mathbb{T}$ *is the theory of right* R*-modules, the pure* $\mathbb{T}$*-ideals as in definition* V - 5 *are exactly the* 2*-sided right-pure ideals of* R.

Any pure $\mathbb{T}$-ideal is a 2-sided left-pure ideal by corollary 3 and proposition 5; the converse is true by proposition 9.

Theorem 11.

Let R *be a ring with unit. Consider the set* r-pp(R) *of* 2*-sided right-pure ideals* J *such that, for any* 2*-sided right-pure ideals* I_1, I_2

$$I_1 \cap I_2 \subset J \Rightarrow I_1 \subseteq J \text{ or } I_2 \subseteq J.$$

For any 2*-sided right pure ideal* I *define*

$$\mathcal{O}_I = \{J \in \text{r-pp}(R) \mid I \not\subseteq J\}.$$

The subsets $\mathcal{O}_I$ *constitute a topology on* r-pp(R). *This space is called the right pure spectrum of* R*; it is compact.*

By propositions VI - 10, V - 14, V - 17. ■

Thus for a ring R, we have defined two different spectra : the right pure spectrum of R and dually (working with left-modules) the left pure spectrum of R. These two spectra are generally not homeomorphic; a counterexample will be given in § VII - 4.

§ 5. PURE REPRESENTATION OF A MODULE

If A is a right R-module, theorem V - 19 presents A as the module of global sections of some sheaf ΔA of right R-modules. The results of §§ 2 - 3 produce an easy description ΔA. From this description of ΔA, it follows immediately that ΔR is in fact a sheaf of rings and that ΔA is a sheaf of modules on the sheaf of rings ΔR.

Theorem 12.

Let R be a ring with unit and $\underline{\mathrm{Mod}}_R$ the category of right R-modules. For any 2-sided right pure ideal I of R, define

$$\Delta R(\mathcal{O}_I) = \underline{\mathrm{Mod}}_R(I, I).$$

ΔR is a sheaf of rings on the right pure spectrum of R; R is isomorphic to the ring of global sections of ΔR.

If $\mathcal{U}$ is the formal initial segment of $\underline{\mathrm{Mod}}_R$ generated by I, we have

$$\begin{aligned}\underline{\mathrm{Mod}}_R(I, I) &\cong \mathcal{U}(I, I)\\ &\cong \mathcal{U}(I, \mathcal{U}^* R)\\ &\cong \underline{\mathrm{Mod}}_R(I, \mathcal{U}^* R)\\ &\cong \mathcal{U}_* \mathcal{U}^* R.\end{aligned}$$

as follows from the considerations of proposition 9. Thus ΔR is exactly the sheaf considered in theorem V - 19 and proposition II - 14. The composition of linear endomorphisms makes $\underline{\mathrm{Mod}}_R(I, I)$ into a ring and thus ΔR into a sheaf of rings. ■

Theorem 13.

Let R be a ring with unit and A a right R-module. For any 2-sided right pure ideal I of R define

$$\Delta A(\mathcal{O}_I) = \underline{\mathrm{Mod}}_R(I, A).$$

ΔA is a sheaf of right modules on the sheaf of rings ΔR; A is isomorphic to the module of global sections of ΔA.

If $\mathcal{U}$ is the formal initial segment of $\underline{\mathrm{Mod}}_R$ generated by I, we have

$$\begin{aligned} \underline{\mathrm{Mod}}_R(I, A) &\cong \underline{\mathrm{Mod}}_R(\mathcal{U}_! \, I, A) \\ &\cong \mathcal{U}(I, \mathcal{U}^* A) \\ &\cong \underline{\mathrm{Mod}}_R(I, \mathcal{U}^* A) \\ &\cong \mathcal{U}_* \, \mathcal{U}^* A \end{aligned}$$

as follows from the considerations of proposition 9. Thus ΔA is exactly the sheaf considered in theorem V - 19 and proposition II - 14. Moreover $\underline{\mathrm{Mod}}_R(I, A)$ is a right module on $\underline{\mathrm{Mod}}_R(I, I)$; the scalar multiplication is given by

$$\underline{\mathrm{Mod}}_R(I, A) \times \underline{\mathrm{Mod}}_R(I, I) \to \underline{\mathrm{Mod}}_R(I, A)$$

$$(f, g) \longmapsto f \circ g.$$

■

CHAPTER 7 : PURE REPRESENTATION OF RINGS

In chapter 6, we obtained, from the general theory of formal initial segments, the description of the pure spectra of a ring R and the corresponding representation theorems for R and any R-module. The object of this chapter is twofold : we intend to study more deeply pure ideals and the representation theorems; on the other hand, we want to give a direct treatment of what has been done in chapter 6, i.e. a treatment independant of the theory of Π-ideals. However we insist on the fact that all the results of chapter 6 have been discovered first from the general theory of formal initial segments; the direct algebraic treatment came later.

We work on an arbitrary ring with unit R, not necessarily commutative. When nothing is specified, "module" and "ideal" always mean "right R-module" and "right R-ideal". We denote by $\underline{\mathrm{Mod}}_R$ the category of right R-modules and R-linear mappings. If M and N are two modules, (M, N) denotes the set of linear mappings from M to N.

Several notions of "pure submodule" can be found in the litterature. In the case of a 2-sided ideal, they turn out to be equivalent; this is what we prove in § 1. The definition we adopt is the one wich appears to be most useful in the proofs : a pure ideal of R is a 2-sided ideal I of R such that for every $i \in I$, there exists an element $\varepsilon \in I$ such that $i \cdot \varepsilon = i$. There is clearly a dual notion with ε "unit" on the left. In § 1, we describe also some basic properties of pure ideals and in § 2, we give examples.

A spectrum of a ring R is some topological space associated to the ring R and whose topological properties reflect some aspects of the algebraic structure of R. For example, Grothendieck's spectrum is constructed from the prime ideals of R, Pierce's spectrum is constructed from the idempotents of R, and so on ... . In § 3, we propose a spectrum of R - we call it the pure spectrum of R - constructed from the pure ideals of R; it is always a compact (not necessarily Hausdorff) sober space. As the notion of pure ideal can be defined on the left and on the right, we obtain in fact two different pure spectra of R : a right one and a left one; they are generally not homeomorphic. In § 4, we give some examples and counterexamples.

If X is some topological space, a sheaf of rings on X can be regarded in two ways : to any open subset U of X, we assign a ring F(U) in such a way that certain restriction and glueing conditions are satisfied; or we consider a local homeomorphism $p : F \to X$ such that the family $(p^{-1}(x))_{x \in X}$ is a continuous family of rings. The correspondance between the two definitions comes from the fact that $F(U)$ is isomorphic to the ring of continuous sections of p on the open subset U. The continuous sections of p on the total space X are called the global sections. The ring $p^{-1}(x)$, for $x \in X$, is called the stalk of the sheaf at x.

When a spectrum has been defined for a ring R, one tries generally to construct a sheaf of rings on this spectrum in such a way that R is isomorphic to the ring of global sections of this sheaf. This process is interesting as soon as the stalks of the sheaf have additional properties : in Grothendieck's case, they are local rings; in Pierce's case, they are fields as soon as the ring is von Neumann regular. Thus, for example, the study of a regular ring can be reduced to the study of fields if one accepts to replace a single ring by a sheaf of rings.

In §§ 5 - 6, we propose two different sheaf representations of a ring R on its pure spectrum. The first representation is easily described as a mapping on the open subsets of the spectrum via the rings of endomorphisms of the pure ideals. The second representation has the advantage that the stalks of the sheaf are quotients of the ring R. At the same time we give analogous representation theorems for R-modules. In chapter 8, we shall study the rings for which these representations have nice properties : these are the Gelfand rings; in particular, both representations will coincide for Gelfand rings.

§ 7 is merely a counterexample. We show that Pierce's method for constructing a sheaf representation in terms of "espace étalé" does not work in general when dealing with the pure spectrum. In fact our representations of §§ 5 - 6 both coincide with that of Pierce in the case of regular rings (see chapter 8). But in general the pure spectrum is richer than Pierce's spectrum and the representation theorem splits into two different results.

Finally in §§ 8 - 9, we look at what happens to pure ideals and the pure spectrum when we let the ring R vary. We find that finite products of rings commute with the construction of pure spectra. On the other hand, we need the commutativity of the ring to prove that a ring homomorphism induces a continuous mapping between the corresponding spectra.

§ 1. PURE IDEALS OF A RING

Let R be a ring with unit. We define a (right) pure ideal an give equivalent definitions. We prove some properties of pure ideals.

Definition 1.

A (right) pure ideal of R *is a 2-sided ideal* I *of* R *such that for every* $i \in I$, *there exists an element* $\varepsilon \in I$ *such that* $i\,\varepsilon = i$.

Again we use the convention that, when nothing is specified, "pure" means "right pure". In the same way, $\mathrm{Ann}\; i = \{r \mid i\, r = 0\}$ is the right annihilator of $i \in R$. Several aspects of the following proposition are well known (cfr. [5]).

Proposition 2.

The following conditions are equivalent for a 2-sided ideal I *of* R

(1) I *is pure*

(2) $\forall\, i \in I \quad \exists\, \varepsilon \in I \quad i = i\,\varepsilon$

(3) $\forall\, i_1, \ldots, i_n \in I \quad \exists\, \varepsilon \in I \quad \forall\, k \quad i_k\, \varepsilon = i_k$

(4) $\forall\, A \in \underline{\mathrm{Mod}}_R \quad A \otimes I \cong A\,.\,I$

(5) $\forall\, A \in \underline{\mathrm{Mod}}_R \quad A \otimes I \to A \otimes R$ *is injective*

(6) $R/_I$ *is a left flat module*

(7) *for any ideal* J, $J \cap I = J\,.\,I$

(8) $\forall\, i \in I \quad I + \mathrm{Ann}\; i = R.$

(1) $\Leftrightarrow$ (2) by definition 1. Clearly, (3) $\Rightarrow$ (2) is obvious. We will prove (2) $\Rightarrow$ (3) by induction on n. (3) is valid when $n = 1$ (by (2)). Now suppose that (3) is valid for n and let $i_1, \ldots, i_{n+1}$ be n+1 elements in I. Choose

$\varepsilon \in I$ such that $i_{n+1}\,\varepsilon = i_{n+1}$,

$\varphi \in I$ such that for $k = 1, \ldots, n \quad (i_k - i_k\,\varepsilon)\varphi = (i_k - i_k\,\varepsilon)$.

This implies that

$$\begin{aligned} i_{n+1}\,(\varepsilon + \varphi - \varepsilon\,\varphi) &= i_{n+1}\,\varepsilon + i_{n+1}\,\varphi - i_{n+1}\,\varepsilon\,\varphi \\ &= i_{n+1} + i_{n+1}\,\varphi - i_{n+1}\,\varphi \\ &= i_{n+1} \end{aligned}$$

and for $k = 1, \ldots, n$

$$i_k(\varepsilon + \varphi - \varepsilon\varphi) = i_k\varepsilon + i_k\varphi - i_k\varepsilon\varphi$$
$$= i_k\varepsilon + (i_k - i_k\varepsilon)\varphi$$
$$= i_k\varepsilon + i_k - i_k\varepsilon$$
$$= i_k.$$

Thus $\varepsilon + \varphi - \varepsilon\varphi \in I$ satisfies (3) at the level $n + 1$.

To prove (3) $\Rightarrow$ (4), observe there is a canonical linear mapping

$$A \otimes I \to A \,.\, I \;;\quad a \otimes i \mapsto a\, i$$

and (4) must be understood as the fact that this mapping is an isomorphism. It is clearly surjective as any $\sum_k a_k\, i_k \in A.\, I$ is the image of $\sum_k a_k \otimes i_k \in A \otimes I$. We will now show that this mapping is injective. If $\sum_k a_k \otimes i_k$ is sended to 0, i.e. if $\sum_k a_k\, i_k = 0$, choose $\varepsilon \in I$ such that for any k, $i_k = i_k\,\varepsilon$

Then, $\sum_k a_k \otimes i_k = \sum_k (a_k \otimes i_k\,\varepsilon)$

$$= \sum_k (a_k\, i_k \otimes \varepsilon)$$
$$= (\sum_k a_k\, i_k \otimes \varepsilon$$
$$= 0 \otimes \varepsilon$$
$$= 0.$$

This proves the injectivity and finally the isomorphism $A \otimes I \cong A \,.\, I$.

Conversely suppose (4) to be satisfied and for any $i \in I$, consider the exact sequence of modules

$$0 \longrightarrow i\,R \longrightarrow R \longrightarrow \frac{R}{i\,R} \longrightarrow 0.$$

Tensoring with I, we obtain an exact sequence

$$i\,R \otimes I \longrightarrow R \otimes I \longrightarrow \frac{R}{i\,R} \otimes I \longrightarrow 0$$

or, using (4)

$$i\,R\,I \longrightarrow R\,I \longrightarrow \frac{R}{i\,R}\, I \longrightarrow 0.$$

But I is 2-sided, thus we obtain

$$i\,I \longrightarrow I \longrightarrow \frac{I}{i\,R} \longrightarrow 0.$$

Now $i \in I$ is sended to 0 in $\frac{I}{i\,R}$, thus by exactness of the sequence, it is the image of some element $\sum_k i\, i_k \in i\, I$. So

$$i = \sum_k i\, i_k = i(\sum_k i_k)$$

and $\sum_k i_k \in I$. So I satisfies (2) and we proved the implication (4) $\Rightarrow$ (2).

The equivalence (4) $\Leftrightarrow$ (5) is easy. It suffices to consider the factorization, valid for any module M,

$$M \otimes I \twoheadrightarrow M\,.\,I \rightarrowtail M\,.\,R = M.$$

The first mapping is thus an isomorphism if and only if it - or equivalently the composite - is injective. But this is the equivalence (4) $\Leftrightarrow$ 5. Thus we have already proved (1) $\Leftrightarrow$ (2) $\Leftrightarrow$ (3) $\Leftrightarrow$ (4) $\Leftrightarrow$ (5). It should be pointed out that (5) is simply Cohn's definition of a pure left-submodule $I \rightarrowtail R$ (cfr. [5]).

We will now prove (2) $\Rightarrow$ (7). If J is right sided, $J\,I \subseteq J$ and since I is left sided, $J\,I \subseteq I$; thus $J\,I \subseteq J \cap I$. Now take $i \in J \cap I$ and choose $\varepsilon \in I$ such that $i = i\,\varepsilon$. We have $i \in J$ and $\varepsilon \in I$, thus $i\,\varepsilon = i \in I\,.\,J$. Conversely suppose (7) to be satisfied and choose $i \in I$. From the equality

$$i\,R = i\,R \cap I = i\,R\,.\,I = i\,I,$$

we deduce that $i = i\,.\,1 \in i\,R$ is in $i\,I$, thus

$$i = \sum_k i\, i_k = i(\sum_k i_k)$$

with $i_k \in I$ and thus $\sum_k i_k \in I$; this is (2).

Now we must prove that (6) is equivalent to the other conditions. Suppose I is pure. $R/_I$ is a left flat module if for any right ideal J, the morphism $J \otimes R/_I \to R \otimes R/_I$ is injective. Consider the exact sequence of left modules

$$0 \longrightarrow I \longrightarrow R \longrightarrow R/_I \longrightarrow 0.$$

Tensoring by J, we obtain an exact sequence

$$J \otimes I \longrightarrow J \otimes R \longrightarrow J \otimes \frac{R}{I} \longrightarrow 0$$

or equivalently

$$0 \longrightarrow J \cap I \longrightarrow J \longrightarrow J \otimes R/_I \longrightarrow 0.$$

This proves the isomorphism

$$J \otimes R/_I \cong \frac{J}{J \cap I}.$$

Finally we need to show that

$$\frac{J}{J \cap I} \longrightarrow \frac{R}{I},$$

is injective, which is obvious.

Now suppose that $R/_I$ is a left flat module and choose $i \in I$. Tensoring the injection $i\,R \rightarrowtail R$ with the left flat module $R/_I$, we obtain an injection

$$i\,R \otimes R/_I \rightarrowtail R \otimes R/_I \cong R/_I.$$

Now any generator $i\,r \otimes s$ of $iR \otimes R/_I$ is sended to $i\,r\,s \in I$ in $R/_I$, thus to 0. By injectivity, $i\,R \otimes R/_I = (0)$. Consider the exact sequence

$$0 \longrightarrow I \longrightarrow R \longrightarrow R/_I \longrightarrow 0.$$

Tensoring with $i\,R$ we obtain an exact sequence

$$i\,R \otimes I \longrightarrow i\,R \otimes R \longrightarrow i\,R \otimes R/_I \longrightarrow 0$$

or equivalently

$$i\,R \otimes I \longrightarrow i\,R \longrightarrow 0.$$

This proves that the mapping $i\,R \otimes I \to i\,R$ is surjective; thus $i = i.1 \in iR$ is the image of $\sum_k i\,r_k \otimes i_k$; so

$$i = \sum_k i\,r_k\,i_k = i\,(\sum_k r_k\,i_k)$$

and $\sum_k r_k\,i_k \in I$. This proves (2).

Finally we prove the equivalence (2) $\Leftrightarrow$ (8). If I satisfies (2) and $i \in I$, the annihilator of i is the ideal defined by

$$\text{Ann}\ i = \{r \in R \mid i\,r = 0\}.$$

Choose $\varepsilon \in I$ such that $i\,\varepsilon = i$; this implies $i(1 - \varepsilon) = 0$ and thus $1 - \varepsilon \in \text{Ann}\ i$. Therefore we have

$$I + \text{Ann}\ i = R$$

because $\varepsilon \in I$ and $1 - \varepsilon \in \text{Ann}\ i$, thus

$$1 = (1 - \varepsilon) + \varepsilon \in I + \text{Ann}\ i.$$

Conversely if

$$I + \text{Ann } i = R$$

then for $i \in I$, we can write

$$\varepsilon + r = 1$$

where $\varepsilon \in I$ and $r \in \text{Ann } i$. Multiplying both sides by i, we get

$$i\,\varepsilon + i\,r = i.$$

But $i\,r = 0$ since $i\,r \in \text{Ann } i$. Thus $i = i\,\varepsilon$. ■

<u>Proposition 3.</u>

Let I *be a pure ideal of* R *and* $r \in R$. *Then*

$$r \in I \iff I + \text{Ann } r = R.$$

One implication is simply (8) in proposition 2. Now if

$$I + \text{Ann } r = R,$$

write $1 = \varepsilon + \varphi$ with $\varepsilon \in I$ and $\varphi \in \text{Ann } r$. Multiplying by r, we obtain

$$r = r\,\varepsilon + r\,\varphi = r\,\varepsilon + 0 = r\,\varepsilon \in I.$$ ■

<u>Proposition 4.</u>

Any (right) pure ideal is a left flat module.

For any injection $S \rightarrowtail A$ of (right) modules, we must prove the injectivity of $S \otimes I \to A \otimes I$; but this is simply the inclusion $S\,I \hookrightarrow A\,I$ (proposition 4). ■

<u>Proposition 5.</u>

If R *is a commutative ring with unit and* I *a pure ideal, then the ring* (I, I) *of linear endomorphisms of* I *is commutative.*

Choose f, g two linear endomorphisms of I. For any $i \in I$, choose $\varepsilon \in I$ such that $i\,\varepsilon = i$.

$$(f \circ g)(i) = f(g(i\,\varepsilon)) = f(i\,g(\varepsilon)) = g(\varepsilon)\,f(i)$$
$$(g \circ f)(i) = g(f(i\,\varepsilon)) = g(\varepsilon\,f(i)) = g(\varepsilon)\,f(i).$$ ■

<u>Proposition 6.</u>

Let I *be a pure ideal and* J_1, J_2 *two ideals. Then*

$$\left.\begin{array}{l} I + J_1 = I + J_2 \\ I \cap J_1 = I \cap J_2 \end{array}\right\} \Rightarrow J_1 = J_2.$$

Take $a \in J_1 \subseteq I + J_1 = I + J_2$. We can write $a = i + j$ with $i \in I$, $j \in J_2$. Choose ε such that $i\,\varepsilon = i$.

$$a\,\varepsilon = (i + j)\varepsilon = i\,\varepsilon + j\,\varepsilon = i + j\,\varepsilon.$$

Therefore

$$i = a\,\varepsilon - j\,\varepsilon \in I \cap J_1 = I \cap J_2 \subseteq J_2.$$

So $a = i + j \in J_2$ and thus $J_1 \subseteq J_2$. Conversely $J_2 \subseteq J_1$. ■

Proposition 7.

(0) *and* R *are pure ideals of* R.

Any sum and any finite intersection of pure ideals is a pure ideal.

0 is a unit in (0) and 1 is a unit in R so (0) and R are trivially pure.

Let $(I_k)_{k\in K}$ be a family of pure ideals of R. An element in $\underset{k\in K}{+} I_k$ has the form $\sum_{\ell=1}^{n} i_\ell$ where $i_\ell \in I_{k_\ell}$. We will show, by induction on n, that there exists some $\varepsilon \in \overset{n}{\underset{\ell=1}{+}} I_{k_\ell}$ such that $(\sum_{\ell=1}^{n} i_\ell)\varepsilon = \sum_{\ell=1}^{n} i_\ell$. If $n = 1$, $i_1 \in I_{k_1}$ and thus there is $\varepsilon \in I_{k_1}$ such that $i_1 \cdot \varepsilon = i_1$. Now suppose the result is true for n. To prove it for $n + 1$, choose $\varepsilon \in I_{k_{n+1}}$ such that $i_{n+1} \cdot \varepsilon = i_{n+1}$. Consider also

$$\sum_{\ell=1}^{n} i_\ell - i_\ell\,\varepsilon \in \overset{n}{\underset{\ell=1}{+}} I_\ell$$

and choose $\varphi \in \overset{n}{\underset{\ell=1}{+}} I_\ell$ such that

$$(\sum_{\ell=1}^{n} i_\ell - i_\ell\,\varepsilon)\varphi = \sum_{\ell=1}^{n} i_\ell - i_\ell\,\varepsilon.$$

We have $\varepsilon + \varphi - \varepsilon\,\varphi \in \overset{n+1}{\underset{i=1}{+}} I_\ell$ and

$$(\sum_{\ell=1}^{n+1} i_\ell)\,(\varepsilon + \varphi - \varepsilon\varphi)$$

$$= (\sum_{\ell=1}^{n} i_\ell - i_\ell\,\varepsilon)\varphi + \sum_{\ell=1}^{n} i_\ell\,\varepsilon + i_{n+1}\,\varepsilon + i_{n+1}\,\varphi - i_{n+1}\,\varepsilon\,\varphi$$

$$= (\sum_{\ell=1}^{n} i_\ell - \sum_{\ell=1}^{n} i_\ell \,\varepsilon + \sum_{\ell=1}^{n} i_\ell \,\varepsilon + i_{n+1} + i_{n+1}\,\varphi - i_{n+1}\,\varphi$$

$$= \sum_{\ell=1}^{n+1} i_\ell .$$

Now, take I, J pure in R and $i \in I \cap J$. Choose $\varepsilon \in I$ and $\varphi \in J$ such that $i\,\varepsilon = i$, $i\,\varphi = i$. Then $i\,\varepsilon\,\varphi = i$ and $\varepsilon\,\varphi = I\,J = I \cap J$. ∎

Proposition 8.

Any ideal I *contains a largest pure ideal. We call it the pure part of* I*; it is denoted by* $\overset{\circ}{I}$.

$\overset{\circ}{I}$ is simply the sum of all pure ideals contained in I. Such ideals exist (at least (0)) and their sum is still a pure ideal (proposition 7); it is obviously the largest pure ideal contained in I. ∎

Proposition 9.

Let I, J *be two ideals and* $(I_k)_{k \in K}$ *a family of ideals. Then*

$$\overset{\circ}{\widehat{I \cap J}} = \overset{\circ}{I} \cap \overset{\circ}{J},$$

$$\overset{\circ}{\widehat{\underset{k \in K}{+} I_k}} \supseteq \underset{k \in K}{+} \overset{\circ}{I}_k .$$

We have $\overset{\circ}{I} \subseteq I$ and $\overset{\circ}{J} \subseteq J$, thus

$$\overset{\circ}{I} \cap \overset{\circ}{J} \subseteq I \cap J$$

and $\overset{\circ}{I} \cap \overset{\circ}{J}$ is pure by proposition 3. This proves

$$\overset{\circ}{I} \cap \overset{\circ}{J} \subseteq \overset{\circ}{\widehat{I \cap J}} .$$

Conversely $\overset{\circ}{\widehat{I \cap J}} \subseteq I \cap J \subseteq I$ and $\overset{\circ}{\widehat{I \cap J}}$ is pure; this prove $\overset{\circ}{\widehat{I \cap J}} \subseteq \overset{\circ}{I}$ and in the same way, $\overset{\circ}{\widehat{I \cap J}} \subseteq \overset{\circ}{J}$.

Finally

$$\overset{\circ}{\widehat{I \cap J}} \subseteq \overset{\circ}{I} \cap \overset{\circ}{J} .$$

Since $\underset{k \in I}{+} \overset{\circ}{I}_k$ is a pure ideal (proposition 7) contained in $\underset{k \in K}{+} I_k$, the second relation follows immediately from the definition of pure part. ■

An ideal is generally not the intersection of the maximal ideals above it. Moreover, the pure part of an arbitrary intersection of ideals is generally not the intersection of the pure parts of the ideals. However, the following result holds :

Proposition 10.

Let I *be a pure ideal. Then*

$$I = \overset{\circ}{\widehat{\cap M}} = \cap \overset{\circ}{M}$$

where the intersections are over all 2-sided (resp. right) maximal ideals M *containing* I.

The following proof works in both cases of 2-sided or right ideals. $I \subseteq M$ implies $I \subseteq \cap M$ and since I is pure, $I \subseteq \overset{\circ}{\widehat{\cap M}}$. On the other hand, $\cap M \subseteq M$ implies $\overset{\circ}{\widehat{\cap M}} \subseteq \overset{\circ}{M}$ and finally $\overset{\circ}{\widehat{\cap M}} \subseteq \cap \overset{\circ}{M}$. So it suffices to prove the inclusion $\cap \overset{\circ}{M} \subseteq I$.

If $\cap \overset{\circ}{M} \not\subseteq I$, choose $a \in \cap \overset{\circ}{M} \setminus I$.

From proposition 3, we deduce

$$I + \text{Ann } a \neq R.$$

Choose a maximal N containing I + Ann a, and thus I. We have $a \in \cap \overset{\circ}{M}$, thus $a \in \overset{\circ}{N}$ and by proposition 3

$$\overset{\circ}{N} + \text{Ann } a = R$$

which is a contradiction since $\overset{\circ}{N}$ and Ann a are in N. ■

Proposition 11.

Let A *be a module and* I *a pure ideal. Then*

$$A\, I = \{a \in A \mid \exists\, \varepsilon \in I \quad a = a\, \varepsilon\}.$$

Clearly each $a\,\varepsilon$ with $a \in A$, $\varepsilon \in I$ is in A I. Conversely consider $a = \sum_{k=1}^{n} a_k\, i_k \in A\, I$ and choose $\varepsilon \in I$ such that for any k, $i_k\, \varepsilon = i_k$.

$$a = \sum_k a_k\, i_k = \sum_k (a_k\, i_k\, \varepsilon) = (\sum_k a_k\, i_k)\varepsilon = a\, \varepsilon. \quad \blacksquare$$

Proposition 12.

Let A *and* B *be two modules and* I *a pure ideal. Any linear mapping* $A\,I \to B$ *factors through* $B\,I$.

Take $f : A\,I \to B$ a linear mapping. For any $a \in A\,I$ there is $\varepsilon \in I$ such that $a = a\,\varepsilon$ (proposition 11). Therefore :

$$f(a) = f(a\,\varepsilon) = f(a)\,\varepsilon \in B\,I. \quad \blacksquare$$

Proposition 13.

Let A *be a module,* $(S_k)_{k\in K}$, S, T *submodules of* A *and* I *a pure ideal. Then*

$$A\,I \cap (\underset{k\in K}{+} S_k) = \underset{k\in K}{+} (A\,I \cap S_k)$$

$$S \cap (A\,I + T) = (S \cap A\,I) + (S \cap T)$$

$$A\,I + (S \cap T) = (A\,I + S) \cap (A\,I + T)$$

$$S + (A\,I \cap T) = (S + A\,I) \cap (S + T).$$

The inclusions

$$\underset{k\in K}{+} (A\,I \cap S_k) \subseteq A\,I \cap (\underset{k\in K}{+} S_k)$$

$$(S \cap A\,I) + (S \cap T) \subseteq S \cap (A\,I + T)$$

are obvious and valid for any submodules. We will now prove the converse inclusions.

Take $a \in A\,I \cap (\underset{k\in K}{+} S_k)$. By proposition 11, we can write $a = a\,\varepsilon$ with $\varepsilon \in I$ and $a = s_1 + \dots + s_n$ with $s_\ell \in S_{k_\ell}$. Therefore,

$$a = a\varepsilon = s_1\,\varepsilon + \dots + s_n\,\varepsilon$$

with $s_\ell\,\varepsilon \in A\,I \cap S_{k_\ell}$.

Now take $s \in S \cap (A\,I + T)$. By proposition 11, we can write $s = a\,i + t$ with $a \in A$, $i \in I$, $t \in T$. Choose $\varepsilon \in I$ such that $i\,\varepsilon = i$. This implies

$$s\,\varepsilon = a\,i\,\varepsilon + t\,\varepsilon = a\,i + t\,\varepsilon$$

and thus

$$s = s\,\varepsilon + t - t\,\varepsilon.$$

But $s\,\varepsilon \in A\,I \cap S$ and $t - t\,\varepsilon = s - s\,\varepsilon \in S \cap T$.

The last two relations can be formally deduced from the preceding ones, without going back to the definition of a pure ideal. Indeed, for the third relation, we have

$$\begin{aligned}(A\,I + S) \cap (A\,I + T) &= ((A\,I + S) \cap A\,I) + ((A\,I + S) \cap T)\\ &= A\,I + ((A\,I \cap T) + (S \cap T))\\ &= A\,I + (S \cap T)\end{aligned}$$

and for the last relation

$$\begin{aligned}(S + A\,I) \cap (S + T) &= (S \cap (S + T)) + (A\,I \cap (S + T))\\ &= S + ((A\,I \cap S) + (A\,I \cap T))\\ &= S + (A\,I \cap T).\end{aligned}$$

■

<u>Proposition 14</u>.

Let I *be a pure ideal and* A *a module*

$$\complement A\,I \underset{\text{def}}{=} \{a \in A \mid \forall i \in I \quad a\,i = 0\}$$

is the largest submodule of A *whose intersection with* A I *is zero.*

Consider the submodule

$$S = \sum_{A\,I \cap T = (0)} T$$

where all T are submodules of A. Using proposition 13, we have

$$\begin{aligned}A\,I \cap S &= A\,I \cap \Big(\sum_{A\,I \cap T = (0)} T\Big)\\ &= \sum_{A\,I \cap T = (0)} (A\,I \cap T)\\ &= (0).\end{aligned}$$

Thus S is the largest submodule of A whose intersection with A I is zero. We must prove the equality $S = \complement A\,I$.

Take $s \in S$ and $i \in I$. Then,

$$s\, i \in S \cap A\, I = (0)$$

thus $s\, i = 0$. Conversely take $a \in \complement A\, I$: we will show that

$$a\, R \cap A\, I = (0).$$

If $x \in a\, R \cap A\, I$, by proposition 11, we can find $\varepsilon \in I$ such that $x = x\, \varepsilon$; but we can also find $r \in R$ such that $x = a\, r$. Finally $x = x\, \varepsilon = a\, r\, \varepsilon$ with $r\, \varepsilon \in I$; so $x = 0$. But this proves the inclusion $a\, R \subset S$ and thus $a \in S$. Finally $S = \complement A\, I$, which concludes the proof. ∎

Proposition 15.

Let I, J *be two pure ideals,* $(I_k)_{k \in K}$ *a family of pure ideals and* A *a module. Then,*

(1) $A\, \complement I \leqslant \complement A\, I$

(2) $(I \leqslant J) \Rightarrow (\complement A\, I \geqslant \complement A\, J)$

(3) $\complement A\, (\underset{k \in K}{+} I_k) = \bigcap_{k \in K} \complement (A\, I_k)$

(4) $\complement A\, (I \cap J) \geqslant \complement A\, I + \complement A\, J$.

If $r \in \complement I$, then for any $i \in I$, we have $r\, i = 0$ and for any $a \in A$, we have $a\, r\, i = 0$. Hence, $a\, r$ is in $\complement A\, I$. This proves (1).

The second relation is obvious.

To prove the third relation, consider

$$\begin{aligned} \complement A\, (\underset{k \in K}{+} I_k) &= \{a \in A \mid \forall\, i \in \underset{k \in K}{+} I_k,\ a\, i = 0\} \\ &= \{a \in A \mid \forall\, k \in K\ \ \forall\, i \in I_k,\ a\, i = 0\} \\ &= \bigcap_{k \in K} \{a \in A \mid \forall\, i \in I_k\ \ a\, i = 0\} \\ &= \bigcap_{k \in K} \complement A\, I_k. \end{aligned}$$

Finally the second relation implies $\complement A(I \cap J) \geqslant \complement A\, I$ and $\complement A(I \cap J) \geqslant \complement A\, I$; this implies the fourth relation. ∎

Proposition 16.

Let I *be a pure ideal of* R.

$\complement I$ *is the left annihilator of* I*; it is a 2-sided ideal.*

Indeed, by definition

$$\text{L-Ann } I = \{r \in R \mid \forall i \in I \quad r\,i = 0\} = \complement I.$$

By definition, $\complement I$ is a right ideal (proposition 14) and L-Ann I is a left ideal (obvious); thus $\complement I$ is a 2-sided ideal. ∎

Proposition 17.

Let I *be a pure ideal of* R.
There is a largest pure ideal $\mathring{\complement} I$ *whose intersection with* I *is* (0).

From propositions 14 and 8, it follows that $\mathring{\complement} I$ is simply the pure part of $\complement I$. ∎

Proposition 18.

The assignment

$$I \longmapsto \mathring{\complement}\mathring{\complement} I$$

is a closure operation on the lattice of pure ideals of R.

For any pure ideal I, we will denote $\mathring{\complement}\mathring{\complement} I$ by $\overline{I}$.

a) $\overline{R} = R$ since $\mathring{\complement} R = (0)$.

b) $I \leqslant \overline{I}$ since $I \cap \mathring{\complement} I = (0)$

c) $\overline{\overline{I}} = \overline{I}$ since $\mathring{\complement}\mathring{\complement}\mathring{\complement} I = \mathring{\complement} I$. Indeed, by b) it suffices to prove the inclusion $\mathring{\complement}\mathring{\complement}\mathring{\complement} I \leqslant \mathring{\complement} I$; by definition of $\mathring{\complement} I$, this is equivalent to $\mathring{\complement}\mathring{\complement}\mathring{\complement} I \cap I = (0)$. This last equality holds since $\mathring{\complement}\mathring{\complement}\mathring{\complement} I \cap \mathring{\complement}\mathring{\complement} I = (0)$ and $I \subseteq \mathring{\complement}\mathring{\complement} I$.

d) $I \leqslant J \Rightarrow \overline{I} \leqslant \overline{J}$ since $I \leqslant J$ implies $\mathring{\complement} J \leqslant \mathring{\complement} I$ and thus $\mathring{\complement}\mathring{\complement} I \leqslant \mathring{\complement}\mathring{\complement} J$.

e) $\overline{I \cap J} = \overline{I} \cap \overline{J}$. Indeed, by d) we have $\overline{I \cap J} \leqslant \overline{I}$ and $\overline{I \cap J} \leqslant \overline{J}$. So, $\overline{I \cap J} \leqslant \overline{I} \cap \overline{J}$ and it remains to show that the converse inclusion holds. By definition of $\mathring{\complement}(I \cap J)$ we have

$$\mathring{\complement}(I \cap J) \cap I \cap J = (0).$$

From the definition of $\mathring{\complement} J$ and $\mathring{\complement}\mathring{\complement} J$, we deduce

$$\mathring{\complement}(I \cap J) \cap I \leqslant \mathring{\complement} J$$

$$\mathring{\complement}(I \cap J) \cap I \cap \mathring{\complement}\mathring{\complement} J \leqslant \mathring{\complement} J \cap \mathring{\complement}\mathring{\complement} J = (0).$$

By definition of $\overset{\circ}{C}\, I$ and $\overset{\circ}{C}\overset{\circ}{C}\, I$, this implies

$$\overset{\circ}{C}(I \cap J) \cap \overset{\circ}{C}\overset{\circ}{C}\, J \leqslant \overset{\circ}{C}\, I$$

$$\overset{\circ}{C}(I \cap J) \cap \overset{\circ}{C}\overset{\circ}{C}\, J \cap \overset{\circ}{C}\overset{\circ}{C}\, I \leqslant \overset{\circ}{C}\, I \cap \overset{\circ}{C}\overset{\circ}{C}\, I = (0).$$

Again by definition of $\overset{\circ}{C}\overset{\circ}{C}\,(I \cap J)$, we obtain

$$\overset{\circ}{C}\overset{\circ}{C}\, I \cap \overset{\circ}{C}\overset{\circ}{C}\, J \leqslant \overset{\circ}{C}\overset{\circ}{C}(I \cap J)$$

which is simply

$$\overline{I} \cap \overline{J} \leqslant \overline{I \cap J}.$$

The properties a - b - c - d - e characterize exactly a closure operation. ■

§ 2. EXAMPLES OF PURE IDEALS

In this paragraph, we describe several examples of pure ideals. Some examples are quite trivial (an ideal with a unit), some are more typical (like continuous real functions which are zero at the neighbourhood of some point .). In the non commutative case, we produce examples of ideals which are pure on the left and not on the right and examples of ideals which are both left and right pure.

Example 19.

A 2-sided direct summand is left pure.

Suppose I is 2-sided and J is right-sided, with $I + J = R$ and $I \cap J = (0)$. Consider $1 = \varepsilon + \varphi$ with $\varepsilon \in I$ and $\varphi \in J$. For any $i \in I$,

$$i = \varepsilon\, i + \varphi\, i;$$

but $\varphi\, i \in J\, I \subseteq J \cap I = (0)$; thus $i = \varepsilon\, i$.

Example 20.

A regular ideal (= generated by its central idempotents) is left and right pure.

If $(e_k)_{k \in K}$ is a family of central idempotents in R, the corresponding generated ideal is

$$I = \{ \sum_{i=1}^{n} e_{k_i}\, r_{k_i} \mid r_{k_i} \in R\} = \{ \sum_{i=1}^{n} r_{k_i}\, e_{k_i} \mid r_{k_i} \in R\}.$$

It is a 2-sided ideal. Now for each $k \in K$, $e_k = e_k\, e_k$; thus e_k is a "unit" for itself (on the left and on the right). If e_k, e_ℓ are two central idempotents

$$e_k + e_\ell - e_k e_\ell = e_k + e_\ell - e_\ell e_k \in I$$

is a "unit" on the left and on the right for e_k and e_ℓ; for example

$$\begin{aligned} e_k(e_k + e_\ell - e_k e_\ell) &= e_k e_k + e_k e_\ell - e_k e_k e_\ell \\ &= e_k + e_k e_\ell - e_k e_\ell \\ &= e_k. \end{aligned}$$

Iterating the process, each finite family of central idempotents $e_{k_1}, \ldots, e_{k_n}$ has a "unit" on the right and on the left in I. So if $\sum_{i=1}^{n} e_{k_i} r_{k_i} \in I$, a "unit" on the right and on the left is obtained by choosing ε which is "unit" on the right and on the left for $e_{k_1}, \ldots, e_{k_n}$.

Example 21.

Let X *be a normal topological space and* C *a closed subset of* X. *The continuons functions* $X \to \mathbb{R}$ *which are zero on a neighborhood of* C *constitute a pure ideal in the ring* $C(X, \mathbb{R})$ *of continuous functions* $X \to \mathbb{R}$.

If $f : X \to \mathbb{R}$ is zero on some neighbourhood V of C and $g : X \to \mathbb{R}$ is zero on some neighbourhood W of V, f + g is zero on $V \cap W$ and for any $h : X \to \mathbb{R}$, hf is zero on V. So the set I of continuous functions $X \to \mathbb{R}$ which are zero on a neighbourhood of C is an ideal.

If $f : X \to \mathbb{R}$ is zero on some neighborhood V of C, we may suppose V to be open. As X is normal, we can find a continuous function $\varphi : X \to \mathbb{R}$ such that $\varphi(C) = 0$ and $\varphi(\complement V) = 1$. $W = \varphi^{-1}([-\frac{1}{2}, +\frac{1}{2}])$ is a closed neighborhood of C which is contained in V. Choose ε continuous such that $\varepsilon(W) = 0$ and $\varepsilon(\complement V) = 1$. ε is in I and $f\,\varepsilon = f$ because $\varepsilon(x) = 1$ as soon as $f(x) \neq 0$, (cfr. [8] for the results on normal spaces).

Example 22.

Let V *be an infinite dimensional vector space on some field* K. *Let* R *be the ring of* K*-linear endomorphisms of* V. *The set* I *of* K*-linear endomorphisms of* V *whose image is finite-dimensional is a left and right pure ideal in* R.

If $f : V \to V$ and $g : V \to V$ are two linear endomorphisms, the image of $f + g$ is contained in $\mathrm{Im}\, f + \mathrm{Im}\, g$; thus it is finite-dimensional as soon as $\mathrm{Im}\, f$ and $\mathrm{Im}\, g$ are finite-dimensional. On the other hand, if $h : V \to V$ is any endomorphism and $\mathrm{Im}\, f$ is finite-dimensional, $\mathrm{Im}(f \circ h) \subseteq \mathrm{Im}\, f$ and $\dim(\mathrm{Im}(h \circ f)) \leqslant \dim(\mathrm{Im}\, f)$: so $\mathrm{Im}(f \circ h)$ and $\mathrm{Im}(h \circ f)$ are finite-dimensional. Finally I is a 2-sided ideal.

Now consider $f : V \to V$ with finite dimensional image. Consider a supplementary subspace W of $\mathrm{Im}\, f$. The linear mapping $\varepsilon : V \to V$ which is the identity on $\mathrm{Im}\, f$ and zero on W has the same image as f and $\varepsilon \circ f = f$; thus I is left pure. On the other hand, $f \circ \varepsilon$ is generally different from f because $f \circ \varepsilon(W) = (0)$ where $\varepsilon \circ f(W)$ is not necessarily reduced to (0).

Now consider the kernel "Ker f" of f and a supplementary subspace U of Ker f. Each $v \in V$ can be uniquely written $v = u + k$ with $u \in U$, $k \in \mathrm{Ker}\, f$. Define $\varphi : V \to V$ by $\varphi(v) = u$. The image of φ is U. But f is injective on U since $\mathrm{Ker}\, \varphi \cap U = (0)$; thus $\dim U = \dim f(U) \leqslant \dim f$ and $\dim U$ is finite. So $\varphi \in I$ and

$$\begin{aligned} f(v) &= f(u + k) = f(u) + f(k) = f(u) + 0 \\ &= f(u) = f(\varphi(v)) = (f \circ \varphi)(v) \end{aligned}$$

which proves the equality $f = f \circ \varphi$. So I is right pure. On the other hand, $\varphi \circ f$ is generally different of f since $f(v)$ could - for example - belong to Ker f.

Thus this is an example of an ideal which is left pure and right pure, but the "units" must be choosen differently on the left and on the right.

Example 23.

Let V be a vector space on some field K and $v \neq 0$ some vector in V. Let R be the ring of those linear endomorphisms $f : V \to V$ having v as eigenvector. Let I be the ideal of those endomorphisms $f : V \to V$ having v as eigenvector with eigenvalue 0. I is a right pure ideal which is not left pure.

If $f(v) = \lambda v$ and $g(v) = \mu v$, then $(f \circ g)(v) = \lambda\mu v$ and $(f + g)(v) = (\lambda + \mu)v$. So R is a ring. Now if $f(v) = 0$, $g(v) = 0$ and $h(v) = \lambda v$, one deduces $(f + g)(v) = 0$ and $(h \circ f)(v) = 0 = (f \circ h)(v)$. So I is a 2-sided ideal.

Let $f : V \to V$ be a linear endomorphism with $f(v) = 0$. Let W be a supplementary subspace to the subspace spanned by v. Define $\varepsilon : V \to V$ by $\varepsilon(v) = 0$ and $\varepsilon(w) = w$ if $w \in K$. ε is an element in I and for any $x \in V$, $x = kv + w$ where $k \in K$, $w \in W$ and

$$f(x) = f(kv) + f(w) = f(w) = (f \circ \varepsilon)(x).$$

So $f = f \circ \varepsilon$ and I is right pure.

On the other hand, I is generally not left pure. Indeed, choose f such that $f(w) = v$. For any $\varphi \in I$,

$$(\varphi \circ f)(w) = \varphi(v) = 0 \neq v = f(w)$$

and thus $\varphi \circ f \neq f$. So I is not left pure. ■

§ 3. PURE SPECTRUM OF A RING

In § 1, we described the lattice of (right) pure ideals of a ring R. In proposition 7, we showed that an arbitrary sum and a finite intersection of pure ideals is again a pure ideal. This property is analogous to the one satisfied by the lattice of open subsets of a topological space. In this paragraph, we construct a compact topological space r-Spp(R), called the right pure spectrum of R, whose lattice of open subsets is isomorphic to the lattice of right pure ideals of R. Dually a left pure spectrum ℓ-Spp(R) of R can be constructed; these two spectra are generally not homeomorphic (see example 37, § 4). Several proofs of this paragraph could be shortened using general lattice theory. Again, when nothing is specified, we work with right ideals and right pure ideals.

Definition 24.

A pure ideal I of R is called "purely maximal" if it is maximal in the lattice of proper pure ideals.

Definition 25.

A pure ideal I of R is called "purely prime" if it is proper and if for any pure ideals I_1, I_2 :

$$I_1 \cap I_2 \subseteq I \Rightarrow I_1 \subseteq I \text{ or } I_2 \subseteq I.$$

Proposition 26.

Any purely maximal ideal is purely prime.

Suppose I is purely maximal and I_1, I_2 are pure with $I_1 \cap I_2 \subseteq I$. If $I_1 \not\subseteq I$, then

$$I_1 + I = R$$

and by proposition 13

$$\begin{aligned} I_2 &= I_2 \cap R \\ &= I_2 \cap (I_1 + I) \\ &= (I_2 \cap I_1) + (I_2 \cap I) \\ &\subseteq I + I \\ &= I. \end{aligned}$$

■

Proposition 27.

The pure part of any maximal ideal is purely prime.

Let M be a maximal ideal and $\overset{\circ}{M}$ its pure part. Let $I_1 \cap I_2 \subseteq \overset{\circ}{M}$ with I_1, I_2 pure. If $I_1 \subseteq M$, then $I_1 \subseteq \overset{\circ}{M}$ and the result holds. If $I_1 \not\subseteq M$, then

$$I_1 + M = R$$

and therefore

$$\begin{aligned} I_2 &= I_2 \cap R \\ &= I_2 \cap (I_1 + M) \\ &= (I_2 \cap I_1) + (I_2 \cap M) \\ &\subseteq M + M \\ &\subseteq M \end{aligned}$$

which implies $I_2 \subseteq \overset{\circ}{M}$. ■

The pure part of a maximal ideal need not be purely maximal; a counter-example is given in § 4.

Proposition 28.

Any proper pure ideal is contained in a maximal pure ideal.

Let I be a proper pure ideal of R. Consider the set, ordered by inclusion

$$X = \{J \mid J \text{ proper pure ideal; } J \supseteq I\}.$$

I belongs to X and for any $J \in X$, $1 \notin J$. In particular, any directed union of elements in X is still in X and X is inductively ordered. By Zorn's lemma, we can choose J maximal in X. Any proper pure ideal containing J contains I and is in X; by maximality of J in X, this ideal is simply J. So J is purely maximal.

Proposition 29.

If I *is a pure ideal and* $a \notin I$*, there is a purely prime ideal* J *such that* $I \subseteq J$ *and* $a \notin J$.

By proposition 10, there is a maximal ideal M such that $\mathring{M} \supseteq I$ and $a \notin \mathring{M}$; by proposition 27, $\mathring{M}$ is purely prime. ■

Proposition 30.

Any pure ideal I *is the intersection of the purely prime ideals containing* I.

By propositions 10 and 27. ■

Definition 31.

We denote by p(R) *the lattice of pure ideals of* R *and by* pp(R) *the set of purely prime ideals of* R.

Theorem 32.

For any pure ideal I *of* R *define*

$$\mathcal{O}_I = \{J \in pp(R) \mid I \not\subseteq J\}.$$

The subsets $\mathcal{O}_I$*, with* I *pure, form a topology on the set* pp(R). *Moreover, the assignment*

$$I \longmapsto \mathcal{O}_I$$

is an isomorphism between the lattice p(R) *of pure ideals of* R *and the lattice of open subsets of* pp(R).

$\mathcal{O}_{(o)} = \{J \in pp(R) \mid (o) \not\subseteq J\} = \phi$; thus the empty subset of pp(R) is simply $\mathcal{O}_{(o)}$.

$$\mathcal{O}_R = \{J \in pp(R) \mid R \not\subseteq J\} = pp(R)$$

since a purely prime ideal is proper. So pp(R) is simply $\mathcal{O}_R$.

Consider two pure ideals I_1, I_2.

$$\begin{aligned} \mathcal{O}_{I_1} \cap \mathcal{O}_{I_2} &= \{J \in pp(R) \mid I_1 \not\subseteq J \text{ and } I_2 \not\subseteq J\} \\ &= \{J \in pp(R) \mid I_1 \cap I_2 \not\subseteq J\} \\ &= \mathcal{O}_{I_1 \cap I_2}. \end{aligned}$$

These equalities hold because J is purely prime.

Consider now a family $(I_k)_{k\in K}$ of pure ideals

$$\begin{aligned} \bigcup_{k\in K} \mathcal{O}_{I_k} &= \{J \in pp(R) \mid \exists\, k \in K,\ I_k \not\subseteq J\} \\ &= \{J \in pp(R) \mid \sum_{k\in K} I_k \not\subseteq J\} \\ &= \mathcal{O}_{\sum_{k\in K} I_k} \end{aligned}$$

and $\sum_{k\in K} I_k$ is pure (cfr. proposition 7).

Hence, the subsets $\mathcal{O}_I$ with I pure constitute a topology on pp(R) and the proof above also shows that the assignment $I \longmapsto \mathcal{O}_I$ preserves finite $\wedge$ and arbitrary $\vee$. To conclude the proof, it suffices to show that this assignment is a bijection between the pure ideals and the open subsets of pp(R). Thus we must prove the equivalence

$$I_1 = I_2 \iff \mathcal{O}_{I_1} = \mathcal{O}_{I_2}$$

for any pure ideals I_1, I_2. But $\mathcal{O}_{I_1} = \mathcal{O}_{I_2}$ means that I_1 and I_2 are contained exactly in the same purely prime ideals; thus they are equal by proposition 30. ■

Definition 33.

The pure spectrum of a ring is the set pp(R) *of purely prime ideals of* R *provided with the topology given in theorem* 32*; it is denoted by* Spp(R).

Again, when nothing is specified, everything is defined on the right. Dually a pure spectrum can be defined working with left pure ideals and left purely prime ideals. When some confusion could arise, we will use the more precise notations r-Spp(R) and ℓ-Spp(R). In general, these two spectra are not homeomorphic (cfr. example 37).

Proposition 34.

The pure spectrum of a ring is a compact (not necessarily Hausdorff) space.

Consider a family $\mathcal{O}_{I_k}$ of open subsets of Spp(R) such that $\bigcup_{k \in K} \mathcal{O}_{I_k} = \mathrm{Spp}(R)$. This means that (theorem 32) $\sum_{k \in K}^{+} I_k = R$. Thus $1 \in \sum_{k \in I}^{+} I_k$ and we can choose $k_1, \ldots, k_n \in K$ and $\varepsilon_i \in I_{k_i}$ such that $1 = \varepsilon_1 + \ldots + \varepsilon_n \in I_{k_1} + \ldots + I_{k_n}$. But this implies $R = I_{k_1} + \ldots + I_{k_n}$ and finally $\mathrm{Spp}(R) = \mathcal{O}_{I_{k_1}} \cup \ldots \cup \mathcal{O}_{I_{k_n}}$. ∎

§ 4. EXAMPLES OF PURE SPECTRA

Having defined left and right pure spectra of a ring R, it is natural to ask whether the spectra are Hausforff and whether the left spectrum is always equal or homeomorphic to the right one. The following examples will show that in general, the answer to these questions is no. In example 36, we also produce a 2-sided maximal ideal whose left pure part is not pure maximal.

Example 35.

Example of a non-commutative ring whose left and right pure spectra are equal and homeomorphic to the Sierpinski space.

Let V be a vector space (on any field K) with infinite countable dimension. Let R be the ring of K-linear endomorphisms of V. Let I be the set of those endomorphisms whose image is finite-dimensional. By example 22, we know that I is left and right pure. On the other hand, R has only three 2-sided ideals : (o), I and R (cfr. [4]). Thus the left pure spectrum of R and the right pure spectrum of R are equal and consist of (o), I and R.

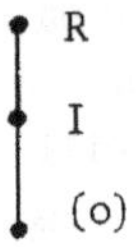

I purely maximal
(o), I purely prime

thus the spectrum is simply

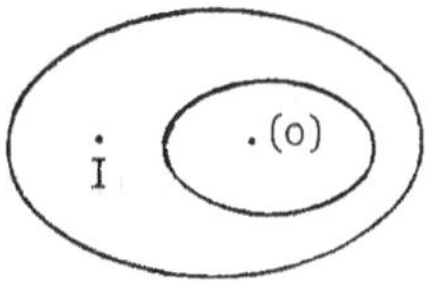

which is the Sierpinski space.

We now recall a proof of the fact that I is the only non trivial pure ideal of R. Let J be a 2-sided ideal and $o \neq f \in J$. Take $g \in R$ with finite dimensional image. Choose a supplementary subspace W of Ker g; g is injective on W and thus W is finite-dimensional. Choose a base $e_1, \ldots, e_n$ of W and a base $e_{n+1}, \ldots, e_k, \ldots$ of Ker g. From $f \neq o$, we deduce that $f(e_m) \neq o$ for some m. We choose another base $f(e_m), e'_2, \ldots, e'_i, \ldots$. For any $p \in \mathbb{N}^*$ and $v \in V$ define two linear mappings r_p and s_v by

$$\begin{cases} r_p(e_p) = e_m \\ r_p(e_i) = 0, \text{ for } i \neq p. \end{cases}$$

$$\begin{cases} s_v(f(e_m)) = v \\ s_v(e'_i) = 0, \text{ for } i \geqslant 2. \end{cases}$$

It follows immediately that for any $i \in \mathbb{N}^*$

$$g(e_i) = \sum_{p=1}^{n} s_{g(e_p)} \circ f \circ r_p(e_i)$$

which proves that

$$g = \sum_{p=1}^{n} s_{g(e_p)} \circ f \circ r_p \in J.$$

This proves that any 2-sided non-zero ideal J contains I.

Now suppose J 2-sided and containing strictly I. This implies the existence of $f \in J$ with infinite-dimensional image. So the image of f has the same dimension as the whole space V and we choose an isomorphism $\varphi : \mathrm{Im}\, f \leftrightarrow V$. Choose also a supplementary subspace W of Ker f : f induces an isomorphism between W and Im f and thus W is also isomorphic to the whole space V. Choose an isomorphism $\psi : V \leftrightarrow W$ and consider $\varphi \circ f \circ \psi \in J$. The composite $\varphi \circ f \circ \psi$ can be factorized in the following way

$$V \overset{\psi}{\longrightarrow} W \overset{f}{\longrightarrow} \operatorname{Im} f \overset{\varphi}{\longrightarrow} V$$

thus it is an isomorphism. So J contains an invertible element and $J = R$. This concludes the proof.

This spectrum clearly is not Hausdorff. It even fails to be T_1.

Example 36.

Example of a non commutative ring whose left and right pure spectra are not equal but are homeomorphic to the $n^{\underline{th}}$ Sierpinski space.

Take some integer $n \geq 2$. Consider the ring R of (upper) triangular $n \times n$ matrices over some field K. If A is a matrix, A_{ij} is the matrix whose (i, j)-entry is the (i, j)-entry of A, while all the other entries are 0. I_{ij} is the matrix whose (i, j)-entry is 1, while all the other entries are 0.

Consider now a left pure ideal J of R and $0 \neq A \in J$. Suppose $a_{ij} \neq 0$. We have

$$I_{ii} \,.\, A \,.\, I_{jj} = A_{ij} \in J.$$

Since J is left pure, there is $E \in J$ with $E \,.\, A_{ij} = A_{ij}$. Computing the (i, j)-entry, we find $e_{ii} \cdot a_{ij} = a_{ij}$, which implies $e_{ii} = 1$. We deduce

$$I_{ii} = E_{ii} = I_{ii} \,.\, E \,.\, I_{ii} \in J.$$

Now if $\ell \leq i$

$$I_{\ell i} = I_{\ell i} \cdot I_{ii} \in J$$

and using the same process as the one applied to A_{ij}, we deduce $I_{\ell\ell} \in J$. Finally each $I_{\ell\ell}$ with $\ell \leq i$ is in J and thus also $I_i = \sum_{\ell=1}^{i} I_{\ell\ell}$. But if B is any matrix such that $b_{k\ell} = 0$ for $k > i$:

$$B = I_i \,.\, B \in J.$$

Finally, if J is a left pure ideal of R and $A \in J$ with $a_{ij} \neq 0$, any matrix B such that $b_{k\ell} = 0$ for $k > i$ is also in J. This proves that the only possible left pure ideals of R are, for $i = 0, 1, \ldots, n$

$L_i = \{B \in R \mid \text{the } k^{\underline{th}} \text{ row of B is zero for } k > i\}$.

In fact, it is obvious that each such L_i is a 2-sided ideal; moreover each L_i is left pure since I_i is a unit on the left for any matrix in L_i. Thus we have described all the left pure ideals of R.

The right pure ideals of R are exactly the left pure ideals of the dual ring of R, i.e. the ring R^* which has the same elements and the same addition as R but multiplication is reversed :

$$A * B \underset{def}{\cong} B.A.$$

In fact, we shall describe a (covariant) isomorphism between R and R^* and this will suffice to prove that the right pure spectrum of R (i.e. the left pure spectrum of $R^* \cong R$) is isomorphic to the left pure spectrum of R.

If A is a n × n matrix, we define A^τ to be the matrix obtained from A by "transposing" around the second diagonal :

$$A^\tau = (a_{n+1-j,\ n+1-i})_{i,j};$$

for example

$$\begin{pmatrix} a & b & c \\ o & d & e \\ o & o & f \end{pmatrix}^\tau = \begin{pmatrix} f & e & c \\ o & d & b \\ o & o & a \end{pmatrix} .$$

The mapping

$$R \longrightarrow R^* \quad ; \quad A \longmapsto A^\tau$$

is such that

$$\begin{aligned} A^{\tau\tau} &= A \\ (A + B)^\tau &= A^\tau + B^\tau \\ (A.B)^\tau &= B^\tau . A^\tau = A^\tau * B^\tau \\ 0^\tau &= 0 \\ 1^\tau &= 1. \end{aligned}$$

Thus, it is a (covariant) isomorphism between R and its dual R^*. Moreover, the nature of the isomorphism $A \longmapsto A^\tau$ shows that the right pure ideals of R are exactly

$$R_i = \{B \in R \mid \text{the } k^{\underline{th}} \text{ colum of B is zero for } k \leqslant i\}$$

for i = 0, 1, ..., n.

Finally both spectra are isomorphic to the nth Sierpinski space but the only ideals which are both left and right pure are (0) and R. Clearly this spectrum is not Hausdorff nor T1.

Consider also the 2-sided ideal R_1, i.e.

$$R_1 = \{B \in R \mid b_{11} = 0\}.$$

This ideal is obviously left maximal, right maximal and 2-sided maximal. But the description of the left pure ideals then clearly shows that the left pure part of R_1, i.e. the largest left pure ideal contained in R_1, is just (0). This produces an example of a maximal ideal whose left pure part is not purely maximal.

Example 37.

Example of a non commutative ring whose left and right pure spectra are not homeomorphic.

Let R be the ring of infinite (indexed by $\mathbb{N} \times \mathbb{N}$) triangular matrices on some field K, with usual addition and multiplication of matrices. The multiplication makes sense because each column of a triangular matrix has only a finite number of non-zero elements.

For any integer i, define

$$L_i = \{B \in R \mid \text{the } k^{\text{th}} \text{ row of } B \text{ is zero for } k > i\}$$
$$R_j = \{B \in R \mid \text{the } k^{\text{th}} \text{ column of } B \text{ is zero for } k \leqslant j\}.$$

It is obvious that L_i and R_j are 2-sided ideals. Moreover L_i is left pure and R_j is right pure. Indeed the matrix

$$(\varepsilon_{k\ell})_{k\ell} \text{ where } \begin{cases} \varepsilon_{\ell\ell} = 1 \text{ if } \ell \leqslant i \\ \varepsilon_{k\ell} = 0 \text{ otherwise} \end{cases}$$

is a left unit for any matrix in L_i. On the other hand, the matrix

$$(\varphi_{k\ell})_{k\ell} \text{ where } \begin{cases} \varphi_{\ell\ell} = 1 \text{ if } \ell \geqslant j \\ \varphi_{k\ell} = 0 \text{ otherwise} \end{cases}$$

is a right unit for any matrix in R_j. It should be pointed out that there are other pure ideals than L_i and R_j : for example the matrices of finite rank constitute a left and right pure ideal of R (the proof given in example 35 works here); as a consequence the intersection of this ideal with each R_j is also right pure. But we don't need these facts to produce our example.

L_1 is the ideal of those matrices which have non zero elements only on the first row; it is a left pure ideal. We shall prove that L_1 is a minimal left pure ideal and in fact is the unique minimal left pure ideal. In other words, we shall prove that any non-zero left pure ideal contains L_1. On the other hand, we shall prove that R does not have any minimal right pure ideal : this will conclude the proof that the left pure spectrum of R is not homeomorphic to its right pure spectrum.

Let J a non-zero left pure ideal of R and $A \in J$ with $a_{ij} \neq 0$. Denote by $I_{k\ell}$ the matrix whose (k,ℓ)-entry is 1, while all the other entries are 0. Then

$$\frac{1}{a_{ij}} I_{1i}.A.I_{jj} = I_{1j} \in J.$$

Because J is left pure, there is $E \in J$ such that $E . I_{1j} = I_{1j}$. Computing the (1, j)-entry, we find $e_{11} . 1 = 1$. But

$$I_{11} = I_{11} . E . I_{11} \in J.$$

Now for any matrix $B \in L_1$,

$$B = I_{11} . B \in J.$$

This proves that J contains L_1.

On the other hand, if J is a non-zero right pure ideal and $A \in J$ with $a_{ij} \neq 0$, then J is not contained in the right pure ideal R_j. This implies that J is not minimal among the non-zero right pure ideals.

Finally the left pure spectrum of R has a smallest non-empty open subset and the right pure spectrum of R has no minimal non-empty open subset. Thus these spectra are not homeomorphic.

§ 5. FIRST REPRESENTATION THEOREM

In this paragraph, we describe for a ring R, a sheaf ΔR of rings on the pure spectrum of R; R is isomorphic to the ring of global sections of this sheaf ΔR. For any R-module A, we also describe a sheaf ΔA of modules on the sheaf of rings ΔR and again A is isomorphic to the module of global sections of ΔA. Again, everything is defined on the right.

Theorem 38.

Let R be a ring and Spp(R) its pure spectrum. For any pure ideal I, the assignment

$$\mathcal{O}_I \mapsto (I, I) \equiv \Delta R(I)$$

defines a sheaf ΔR of rings whose ring of global sections is isomorphic to R. If R is commutative, ΔR is a sheaf of commutative rings.

In order to have a presheaf, we need to define a restriction map $(I, I) \to (J, J)$ when $\mathcal{O}_J \subseteq \mathcal{O}_I$, i.e. when $J \subseteq I$. By proposition 12, this is just the usual restriction of a linear mapping $f : I \to I$ to the submodule J. Now the linear endomorphisms of I form a ring and the restriction mapping $(I, I) \to (J, J)$ is obviously a ring homomorphism. So we have already defined a presheaf ΔR of rings.

This presheaf is separated. Indeed, consider $I = \underset{k\in K}{+} I_k$ in p(R) and $f, g \in (I, I)$ such that for all k, $f|_{I_k} = g|_{I_k}$. Any element $i \in I$ can be written in the form

$$i = i_1 + \ldots + i_n \; ; \; i_k \in I_k$$

and thus

$$\begin{aligned} f(i) &= f(i_1) + \ldots + f(i_n) \\ &= g(i_1) + \ldots + g(i_n) \\ &= g(i). \end{aligned}$$

This shows that $f = g$ and ΔR is separated.

Finally ΔR is a sheaf. Indeed, consider $I = \underset{k\in K}{+} I_k$ in p(R) and $f_k \in (I_k, I_k)$ such that $f_k|_{I_\ell} = f_\ell|_{I_k}$. First we will show that there is no loss of generality in assuming that the family $(I_k)_{k\in K}$ of pure ideals is stable under finite sums or equivalently under binary sums. Indeed, consider $f_k : I_k \to I_k$ and $f_\ell : I_\ell \to I_\ell$ which coincide on $I_k \cap I_k$. Any element i in $I_k + I_\ell$ can be expressed in the form

$$i = i_k + i_\ell \ ; \ i_k \in I_k \ ; \ i_\ell \in I_\ell .$$

Then, we simply define $f : I_k + I_\ell \to I_k + I_\ell$ by

$$f(i) = f_k(i_k) + f_\ell(i_\ell).$$

There is no ambiguity in the definition for if we have two such decompositions of i

$$i = i_k + i_\ell = i'_k + i'_\ell$$

then we deduce

$$i_k - i'_k = i'_\ell - i_\ell \in I_k \cap I_\ell$$

and therefore

$$f_k(i_k - i'_k) = f_\ell(i'_\ell - i_\ell),$$

or in other words

$$f_k(i_k) + f_\ell(i_\ell) = f_k(i'_k) + f_\ell(i'_\ell).$$

So f is correctly defined and is obviously a linear extension of f_k and f_ℓ. Moreover if I_m is any pure ideal in the family (cfr. proposition 13).

$$I_m \cap (I_k + I_\ell) = (I_m \cap I_k) + (I_m \cap I_\ell).$$

So, if $i \in I_m \cap (I_k + I_\ell)$, i can be written in the form

$$i = i_k + i_\ell \ ; \ i_k \in I_m \cap I_k \ ; \ i_\ell \in I_m \cap I_\ell$$

and therefore

$$f(i) = f_k(i_k) + f_\ell(i_\ell) = f_m(i_k) + f_m(i_\ell) = f_m(i).$$

So we may assume the family $(I_k)_{k\in K}$ to be stable under finite sums. Now, if $i \in \underset{k\in K}{+} I_k$, we can write

$$i = i_1 + \dots + i_n \ ; \ i_k \in I_k .$$

Thus i is in a finite sum of I_k's and therefore, by the first part of the proof, we can suppose that i is in some I_ℓ. So we define $f(i) = f_\ell(i)$ and there is no ambiguity in the definition of $f : \underset{k\in K}{+} I_k \to \underset{k\in K}{+} I_k$ because two different f_ℓ agree on i as soon as $f_\ell(i)$ makes sense. Finally f is obviously a linear mapping extending each f_k and ΔR is a sheaf of rings.

The ring of global sections of ΔR is just $\Delta R(R) \cong (R, R)$ which is isomorphic to R itself : the isomorphism sends on element $r \in R$ to the left multiplication by r

$$R \to R \; ; \; s \mapsto r\, s$$

which is however a linear mapping of right R-modules. Conversely, each linear mapping of right R-modules $f : R \to R$ is simply the left multiplication by f(1) :

$$f(s) = f(1 \, . \, s) = f(1) \, . \, s.$$

Now if R is commutative, each ring (I, I) is commutative by proposition 5. ∎

If J is some purely prime ideal of R, the stalk of ΔR at J is the inductive limit of the $\Delta R(I)$ over all $\mathcal{O}_I$ containing J, i.e.

$$(\Delta R)_J = \varinjlim_{I \not\subseteq J} (I, I).$$

We do not know what this stalk would look like in general, nor what its properties could be. However, in the case of Gelfand rings (chapter 8), we shall prove that the stalk $(\Delta R)_J$ is just the quotient $R/_J$.

<u>Theorem 39.</u>

Let R *be a ring,* Spp(R) *its pure spectrum and* A *some* R-*module.*
For any pure ideal I *the assignment*

$$\mathcal{O}_I \mapsto (I, A) \equiv \Delta A(I)$$

defines a sheaf ΔA *of modules on the sheaf of rings* ΔR. A *is isomorphic to the module of global sections of* ΔA.

The group of linear mappings from I to A can be made into a right R-module by the rule

$$(f \, . \, r)(i) = f(r\, i)$$

for $f \in (I, A)$, $r \in R$ and $i \in I$. This makes sense because I is also left sided and thus $r\, i \in I$. Moreover, (I, A) can be made into a (I, I)-module by the multiplication

$$(I, A) \times (I, I) \to (I, A) \; ; \; (f, \varphi) \mapsto f \circ \varphi.$$

The assignment $\mathcal{O}_I \mapsto (I, A)$ defines a presheaf of modules on the sheaf of rings ΔR. Indeed, if $J \leqslant I$, any linear mapping $f : I \to A$ restricts to a linear mapping $f|_J : J \to A$. This produces a restriction mapping $(I, A) \to (J, A)$. Now, if $\varphi \in (I, I)$ is some element of $\Delta R(I)$,

$$(f \circ \varphi)\big|_J = f\big|_J \circ \varphi\big|_J$$

which implies that the restriction mapping $(I, A) \to (J, A)$ is linear. Thus ΔA is a presheaf of modules on ΔR.

We may verify by methods similar to those of Theorem 38 that ΔA is a sheaf.

The module of global sections of ΔA is just $\Delta A(R) = (R, A)$ which is isomorphic to A. The isomorphism sends $a \in A$ to

$$R \to A \; ; \; r \mapsto a\,r$$

and conversely each linear mapping $f : R \to A$ is of this form with $a = f(1)$:

$$f(r) = f(1 \,.\, r) = f(1) \,.\, r. \qquad \blacksquare$$

§ 6. SECOND REPRESENTATION THEOREM

The topic of this paragraph is similar to that of paragraph 5 : we present a ring R as the ring of global sections of some sheaf ∇R on Spp(R) and we show that any R-module A can be presented as the module of global sections of some sheaf ∇A of modules on the sheaf of rings ∇R. The sheaf representations proposed in this paragraph differ generally of those described in paragraph 5; however in the case of Gelfand rings (chapter 8) they will coincide.
Whereas the sheaf ΔR was easily described in terms of rings of endomorphisms, ∇R is defined in two steps : we construct first a presheaf by means of the pseudo complements $\complement I$ of the pure ideals (proposition 14) and we define ∇R to be the associated sheaf. But in the case of ∇R we are able to give some information on the stalks : they are quotient rings of R. Again everything is defined on the right.

Proposition 40.

Let R be a ring. For any pure ideal I the assignment

$$\mathcal{O}_I \mapsto R/\complement I$$

defines a presheaf of rings on the pure spectrum of R.

In proposition 16, we verified that

$$\complement I = \{r \in R \mid \forall i \in I,\ r\,i = 0\}$$

is a 2-sided ideal of R. In particular, $R/_{\complement I}$ is a ring.

If $J \subseteq I$ are pure ideals, then clearly $\complement J \supseteq \complement I$ and we deduce a ring homomorphism

$$R/_{\complement I} \to R/_{\complement J}.$$

This is the restriction map, wich completes the definition of the presheaf of rings. ■

Proposition 41.

For a ring R, *the presheaf defined by the assignment*

$$O_I \mapsto R/_{\complement I}$$

is a separated presheaf.

Consider $I = \underset{k \in I}{+} I_k$ in p(R). We must show that for all $[r], [s] \in R/_{\complement I}$ whose restrictions to any $R/_{\complement I_k}$ coincide, we have $[r] = [s]$. Or equivalently, that $[r] \in R/_{\complement I}$ is zero as soon as each of its restrictions to $R/_{\complement I_k}$ is zero. In other words, we must prove that for any $r \in R$, we have

$$(\forall k \in K \quad r \in \complement I_k) \Rightarrow (r \in \complement I).$$

This follows immediately from proposition 15 :

$$\complement I = \complement\,(\underset{k \in K}{+} I_k) = \bigcap_{k \in K} \complement I_k.$$ ■

Theorem 42.

The sheaf ∇R *associated to the presheaf defined by*

$$O_I \mapsto R/_{\complement I}$$

is a sheaf of rings on Spp(R)*; its ring of global sections is isomorphic to* R.

We recall the definition of the sheaf ∇R associated to the separated presheaf of proposition 40 (cfr. [9]). For a pure ideal I, consider all the hereditary pure coverings of I, i.e. all the families $(I_k)_{k \in K}$ where I_k is pure, $\underset{k \in K}{+} I_k = I$ and such that with any I_k, the family contains any pure ideal smaller than I_k. Now if $(I_k)_{k\in K}$ and $(J_\ell)_{\ell\in L}$ are two such hereditary pure coverings of I, the family $(I_k \cap J_\ell)_{(k,\ell) \in K\times L}$ is again an hereditary pure covering of I and is contained in each of the families $(I_k)_{k\in K}$, $(J_\ell)_{\ell\in L}$. Indeed, $I_k \cap J_\ell$ is pure (proposition 7); now if I is pure and $I \subseteq I_k \cap J_\ell$ then $I \subseteq I_k$ and $I \subseteq J_\ell$ and thus $I = I_{k'} = J_{\ell'} = I_{k'} \cap J_{\ell'}$ for some $k' \in K$ and $\ell' \in L$. Finally,

$$\begin{aligned} \underset{k,\ell}{+} (I_k \cap J_\ell) &= \underset{k}{+} (\underset{\ell}{+} (I_k \cap J_\ell)) \\ &= \underset{k}{+} (I_k \cap (\underset{\ell}{+} J_\ell)) \\ &= \underset{k}{+} I_k \cap I \\ &= \underset{k}{+} I_k \\ &= I. \end{aligned}$$

Now if $(I_k)_{k\in K}$ is an hereditary pure covering of I, a compatible family of elements in $R/C\ I_k$ is a family $([r_k])_{k\in K}$ of elements $[r_k] \in R/C\ I_k$ such that for any k and ℓ in K, the restrictions of $[r_k]$ and $[r_\ell]$ coincide in $R/C\ (I_k \cap I_\ell)$. Two compatible families of elements on two arbitrary hereditary pure coverings of I are called equivalent if they coincide on some common smaller hereditary pure covering. Now, $\nabla R(I)$ is just the quotient by this equivalence relation of the set of all compatible families as described above. If $\left([r_k] \in R/C\ I_k\right)_{k\in K}$ and $\left([s_\ell] \in R/C\ I_\ell\right)_{\ell\in L}$ are two compatible families, they are equivalent to the families $\left([r_k] \in R/C\ (I_k \cap I_\ell)\right)_{(k,\ell) \in K\times L}$ and $\left([s_\ell] \in R/C\ (I_k \cap I_\ell)\right)_{(k,\ell) \in K\times L}$ and these two last families can be added or multiplied component by component. This describes the ring structure of $\nabla R(I)$.

Now if $(I_k)_{k\in K}$ is an hereditary pure covering of I and $J \subseteq I$ in p(R), we deduce by proposition 13

$$J = J \cap I = J \cap (\underset{k\in K}{+} I_k) = \underset{k\in K}{+} (J \cap I_k)$$

and from this, it follows immediately that $(J \cap I_k)_{k\in I}$ is an hereditary pure covering of J. Thus the assignment

$$\left([r_k] \in R/C\, I_k\right)_{k\in K} \mapsto \left([r_k] \in R/C\, (J \cap I_k)\right)_{k\in K}$$

defines the restriction mapping $\nabla R(I) \to \nabla R(J)$.

There is a trivial ring homomorphism

$$R/C\, I \to \nabla R(I) \; ; \; [r] \mapsto ([r])_{p(I)}$$

where $p(I)$ is just the family of all pure sub-ideals of I. By proposition 41, this mapping is injective.

We must compute the ring of global sections of ∇R, i.e. the ring $\nabla R(R)$. Consider any element $([r_k])_{k\in K}$ in $\nabla R(R)$; thus $[r_k] \in R/C\, I_k$ where $\underset{k\in K}{+} I_k = R$. This implies that

$$1 = \varepsilon_{k_1} + \dots + \varepsilon_{k_n} \; ; \; \varepsilon_{k_i} \in I_{k_i}.$$

We shall consider the element

$$r = r_{k_1} \varepsilon_{k_1} + \dots + r_{k_n} \varepsilon_{k_n} \in R$$

and we shall prove that $[r]$ and $[r_k]$ coincide in each $R/C\, I_k$. This will prove that the injection

$$R = R/(o) = R/C\, R \to \nabla R(R)$$

is also a surjection and thus an isomorphism.

So, we need to prove that

$$\forall\, k \in K \quad r - r_k \in C\, I_k.$$

Let us compute this difference in the following way

$$\begin{aligned} r - r_k &= r_{k_1} \varepsilon_{k_1} + \dots + r_{k_n} \varepsilon_{k_n} - r_k \\ &= r_{k_1} \varepsilon_{k_1} + \dots + r_{k_n} \varepsilon_{k_n} - r_k(\varepsilon_{k_1} + \dots + \varepsilon_{k_n}) \\ &= (r_{k_1} - r_k)\varepsilon_{k_1} + \dots + (r_{k_n} - r_k)\varepsilon_{k_n}. \end{aligned}$$

Choose any $s \in I_k$. We have

$$(r - r_k)s = (r_{k_1} - r_k)\varepsilon_{k_1} s + \dots + (r_{k_n} - r_k)\varepsilon_{k_n} s.$$

For any $\ell = k_1, \ldots, k_n$, the element $\varepsilon_\ell\, s$ is in $I_\ell\, I_k = I_\ell \cap I_k$. Now the family $(r_\ell)_{\ell \in K}$ is a compatible family, thus r_ℓ and r_k coincide in $R/\complement\,(I_\ell \cap I_k)$; in other words, $r_\ell - r_k \in \complement\,(I_\ell \cap I_k)$. So each term of the sum is of the form

$$(r_\ell - r_k)\varepsilon_\ell s \quad \text{with } r_\ell - r_k \in \complement\,(I_\ell \cap I_k) \text{ and } \varepsilon_\ell\, s \in I_\ell \cap I_k.$$

Thus each of these terms is zero and $(r - r_k)\, s = 0$, which implies that $r - r_k$ is in $\complement\, I_k$. ■

Corollary 43.

Any local section of the sheaf ∇R *is locally the restriction of a global section.*

By definition of ∇R, for any element $x \in \nabla R(I)$, there is an hereditary pure covering $I = \underset{k \in K}{+} I_k$ such that the restriction x_k of x to I_k is an element $[r_k]$ in $R/\complement\, I_k$. But an element $[r_k]$ in $R/\complement\, I_k$ is just the restriction of the global element

$$r_k \in R \cong \nabla R(R).$$

This is exactly what the corollary means. ■

Proposition 44.

For any purely prime ideal J *of* R, *the stalk of* ∇R *at* J *is the quotient ring of* R *by the 2-sided ideal* $\cup\, \complement\, I$, *where the union is over all pure ideals* I *not contained in* J.

By proposition 16, $\complement\, I$ and thus $+\, \complement\, I$ are 2-sided ideals. So the quotient $R/+\,\complement\, I$ is a ring. The stalk at J is defined to be the inductive limit

$$\varinjlim_{J \in \mathcal{O}_I} \nabla R(I) = \varinjlim_{I \not\subseteq J} \nabla R(I).$$

In other words, the stalk is just the ring

$$\bigcup_{I \not\subseteq J} \nabla R(I) \Big/ \sim$$

where the equivalence relation identifies an element in $\nabla R(I)$ with any of its restrictions.

Let $x \in \nabla R(I)$ with $I \not\subseteq J$. By definition of ∇R, there is an hereditary pure covering $I = + I_k$ such that x restricted to I_k is some $[r_k]$ in $R/\complement I_k$. But from $I = + I_k \not\subseteq J$, we deduce the existence of some k such that $I_k \not\subseteq J$. Thus this particular I_k appears in the union mentioned above and x is equivalent to $[r_k]$. But $[r_k] \in R/\complement I_k$ is itself the restriction of $r_k \in R$ and certainly $R \not\subseteq J$. Finally, each element in $\nabla R(I)$ is equivalent in the union to some element in $R = \nabla R(R)$. This implies that the stalk is just $R/_\sim$ where the equivalence relation identifies an element r to zero if there is some pure ideal $I \not\subseteq J$ such that [r] is zero in $R/\complement I$, i.e. $r \in \complement I$. Finally, define

$$\tilde{J} = \{r \mid \exists I \text{ pure} \; ; \; I \not\subseteq J \; ; \; r \in \complement I\}$$

$$= \bigcup_{\substack{I \not\subseteq J \\ I \text{ pure}}} \complement I = \sum_{\substack{I \not\subseteq J \\ I \text{ pure}}} \complement I.$$

We have shown that the stalk of ∇R at J is just the ring $R/\tilde{J}$. ■

Proposition 45.

If J is a purely maximal ideal of R, the 2-sided ideal $\tilde{J}$ defined in proposition 44 is contained in J.

Let I be a pure ideal such that $I \not\subseteq J$. Since J is purely maximal, $J + I = R$. Therefore, by proposition 13,

$$\begin{aligned} \complement I &= (\complement I) \cap R \\ &= (\complement I) \cap (I + J) \\ &= (\complement I \cap I) + (\complement I \cap J) \\ &= (\complement^I) \cap J. \end{aligned}$$

This proves the inclusion $\complement I \subseteq J$ and thus $\tilde{J} \subseteq J$. ■

Theorem 46.

Let $\mathcal{R}$ be the disjoint union of the rings $R/\tilde{J}$, where the union ranges over all $J \in \mathrm{Spp}(R)$. We provide $\mathcal{R}$ with the final topology for all the mappings

$$\mathrm{Spp}(R) \to \mathcal{R} \; ; \; J \longmapsto [r] \in R/\tilde{J}$$

for any $r \in R$. The mapping

$$p : \mathcal{R} \to \mathrm{Spp}(R) \; ; \; [r] \in R/\tilde{J} \longmapsto J$$

is a local homeomorphism and R is isomorphic to the ring of continuous sections of p.

For any pure ideal I and any element $r \in R$, the composite

$$\mathcal{O}_I \hookrightarrow \mathrm{Spp}(R) \to \mathcal{R} \; ; \; J \mapsto [r] \in R/\tilde{J}$$

is continuous by definition of the topology on $\mathcal{R}$. Now consider some element $x \in \nabla R(I)$; there is a pure covering $I = \underset{k \in K}{+} I_k$ such that for any k,

$$x\big|_{I_k} = [r_k] \in R/\complement I_k .$$

Consider the mapping

$$\mathcal{O}_I \to \mathcal{R} \; ; \; J \mapsto [x] \in R/\tilde{J}.$$

For any k, the restriction of this mapping to $\mathcal{O}_{I_k}$ is

$$\mathcal{O}_{I_k} \hookrightarrow \mathcal{O}_I \to \mathcal{R} \; ; \; J \mapsto [x] = [r_k] \in R/\tilde{J}$$

and we have already proved the continuity of this mapping. But the $\mathcal{O}_{I_k}$'s form an open covering of $\mathcal{O}_I$ (theorem 32); thus the mapping $\mathcal{O}_I \to \mathcal{R}$ is continuous.

Finally the topology on $\mathcal{R}$ is also the final topology for all the mappings

$$\mathcal{O}_I \to \mathcal{R} \; ; \; J \mapsto [x] \in R/\tilde{J}$$

for any pure ideal I and any $x \in \nabla R(I)$. Therefore theorem 46 is just the "espace étalé" version of our theorem 42 (cfr. [9]). The ring structure on the set of continuous sections of p is defined pointwise from the ring structure of each stalk $R/\tilde{J}$. ∎

We proceed now to construct an analogous sheaf representation ∇A for any R-module A.

<u>Proposition 47</u>.

Let A be a R-module. For any pure ideal I the assignment

$$\mathcal{O}_I \mapsto A/\complement AI$$

defines a presheaf of R-modules on the pure spectrum of R.

If $J \subseteq I$ in p(R), then $\complement AJ \supseteq \complement AI$ and thus there is a quotient linear mapping

$$A/\complement AI \to A/\complement AJ$$

which produces the structure of presheaf. ∎

Proposition 48.

For a R-module A, the presheaf

$$\mathcal{O}_I \mapsto A/\complement\, AI$$

defined in proposition 47 is a separated presheaf.

Consider $I = \underset{k\in K}{+} I_k$ in p(R) and $a \in A$ such that for any k, $a \in \complement\, AI_k$. By proposition 15

$$a \in \bigcap_{k\in K} \complement\, AI_k = \complement\,(\underset{k\in K}{+} \complement\, AI_k) = \complement\, I$$

which proves the separation. ■

Theorem 49.

The sheaf ∇A *associated to the presheaf*

$$\mathcal{O}_I \mapsto A/\complement\, AI$$

of proposition 47 is, on the pure spectrum of R, a sheaf of modules on the sheaf of rings ∇R*; its module of global sections is isomorphic to A.*

An element in $\nabla A(I)$ is thus represented by a compatible family $\left([a_k] \in A/\complement\, AI_k\right)_{k\in K}$ of elements for an hereditary pure covering $(I_k)_{k\in K}$ of I; two such families are equivalent if they coincide on some common smaller hereditary pure covering of I. Now an element in $\nabla R(I)$ is represented by a compatible family $\left([r_\ell] \in R/\complement\, I_\ell\right)_{\ell\in L}$ for an hereditary pure covering $(I_\ell)_{\ell\in L}$ of I. We define an action of the ring $\nabla R(I)$ on the R-module $\nabla A(I)$ by

$$\left[([a_k])_{k\in K}\right] \cdot \left[([r_\ell])_{\ell\in L}\right] = \left[([a_m.r_m])_{m\in M}\right]$$

where $(I_m)_{m\in M}$ is a smaller hereditary pure covering common to $(I_k)_{k\in K}$ and $(I_\ell)_{\ell\in L}$ (for example that described in the beginning of the proof of theorem 42). This definition is clearly compatible with the equivalence relations defining $\nabla A(I)$ and $\nabla R(I)$ but we still need to prove that it gives a structure of $\nabla R(I)$-module on $\nabla A(I)$. The only thing to prove is that for any $m \in M$

$$[a_m] \in A/\complement\, AI_m \text{ and } [r_m] \in R/\complement\, I_m \Rightarrow [a_m.r_m] \in A/\complement\, AI_m.$$

Or in other words

$$a_m \in \complement\, AI_m \text{ or } r_m \in \complement\, I_m \Rightarrow a_m\, r_m \in \complement\, AI_m.$$

The implication

$$a_m \in \complement AI_m \Rightarrow a_m r_m \in \complement AI_m$$

holds because $\complement AI_m$ is a right R-module. The implication

$$r_m \in \complement I_m \Rightarrow a_m r_m \in \complement AI_m$$

holds because $A . \complement I_m \subseteq \complement A.I_m$ (proposition 15).

Consider the R-linear mapping

$$A/_{\complement AI} \to \nabla A(I)$$

which sends $[a]$ on $([a])_{p(I)}$ where $p(I)$ is the set of pure subideals of I. By proposition 48, this mapping is injective. In particular we have an injection

$$A = A/_0 = A/_{\complement AR} \to \nabla A(R).$$

We shall prove that this is also a surjection and thus an isomorphism. Consider an element in $\nabla A(R)$ represented by a family $([a_k])_{k\in K}$ where $(I_k)_{k\in K}$ is an hereditary pure covering of R and $[a_k] \in A/_{\complement AI_k}$. We can write

$$1 = \varepsilon_{k_1} + \dots + \varepsilon_{k_n} \; ; \; \varepsilon_{k_i} \in I_{k_i}.$$

Consider the element

$$a = a_{k_1} \varepsilon_{k_1} + \dots + a_{k_n} \varepsilon_{k_n} \in A,$$

and for any $k \in K$, compute

$$a - a_k = a_{k_1} \varepsilon_{k_1} + \dots + a_{k_n} \varepsilon_{k_n} - a_k$$

$$= a_{k_1} \varepsilon_{k_1} + \dots + a_{k_n} \varepsilon_{k_n} - a_k(\varepsilon_{k_1} + \dots + \varepsilon_{k_n})$$

$$= (a - a_k)\varepsilon_{k_1} + \dots + (a_{k_n} - a_k)\varepsilon_{k_n}.$$

For any $s \in I_k$ and any $\ell = k_1, \dots, k_n$

$$\varepsilon_\ell \, s \in I_\ell \, I_k = I_\ell \cap I_k$$

(cfr. proposition 2). The compatibility of the family $([a_k])_{k\in K}$ implies that $[a_k]$ and $[a_\ell]$ coincide in $A/_{\complement A(I_\ell \cap I_k)}$; in other words,

$$a_\ell - a_k \in \complement A(I_\ell \cap I_k).$$

Finally we have for any ℓ

$$(a_\ell - a_k)\varepsilon_\ell \, s \in A(I_\ell \cap I_k) \cap \complement A(I_\ell \cap I_k) = 0.$$

So for any $s \in I_k$ we have $(a - a_k)\, s = 0$ and thus $a - a_k \in \complement AI_k$. So $[a_k]$ is the restriction at I_k of $a \in A$ and thus the mapping

$$A \to \nabla A(R)$$

is surjective and thus an isomorphism. ■

Corollary 50.

In the sheaf ∇A, *any local section is locally the restriction of a global section.*

Any element $[a]$ in $A/\complement AI$ is the restriction of $a \in A = \nabla A(R)$ and by definition of ∇A, any element of $\nabla A(I)$ is locally in some $A/\complement AI_k$. ■

Proposition 51.

For any purely prime ideal J *of* R *and any module* A, *the stalk of* ∇A *at* J *is the quotient module of* A *by the submodule* $+\complement AI$ *where the sum ranges over all pure ideals* I *not contained in* J.

The stalk at J is the inductive limit

$$\bigcup_{I \not\subseteq J} \nabla A(I) / \sim$$

where the equivalence relation identifies an element in $\nabla A(I)$ with each of its restrictions. But an element in $\nabla A(I)$ is represented by a compatible family $\left\{[a_k] \in A/\complement AI_k\right\}_{k \in K}$ for an hereditary pure covering $(I_k)_{k \in K}$ of I. From $I = \underset{k \in K}{+} I_k \not\subseteq J$, we deduce that for some $k \in K$, $I_k \not\subseteq J$. So the element in $\nabla A(I)$ is equivalent to $[a_k] \in A/\complement AI_k$ and finally to $a_k \in A$. So the stalk at J is just $A/\sim$ where $a \in A$ is equivalent to zero if it is in $\complement AI$ for some pure ideal $I \not\subseteq J$. Thus the stalk at J is the quotient $A/\hat{J}$ where

$$\hat{J} = \bigcup_{\substack{I \not\subseteq J \\ I \text{ pure}}} \complement AI = \underset{\substack{I \not\subseteq J \\ I \text{ pure.}}}{+} \complement AI$$ ■

Proposition 52.

If J *is a purely maximal ideal of* R *and* A *a* R*-module, the submodule* $\hat{J}$ *of* A *defined in proposition* 51 *is contained in* AJ.

If I is pure and $I \not\subseteq J$, we have $I + J = R$ by maximality of J. Therefore by proposition 13

$$\begin{aligned}\complement\, AI &= \complement\, AI \cap A \\ &= \complement\, AI \cap A(I + J) \\ &= \complement\, AI \cap (AI + AJ) \\ &= (\complement\, AI \cap AI) + (\complement\, AI \cap AJ) \\ &= \complement\, AI \cap AJ.\end{aligned}$$

So $\complement\, AI \subset AJ$ and thus $\hat{J} \subseteq AJ$. ■

Theorem 53.

Let A *be a* R-*module. Let* $\mathcal{A}$ *be the disjoint union of the* R-*modules* $A/_{\hat{J}}$, *where the union ranges over all* $J \in Spp(R)$. *We provide* $\mathcal{A}$ *with the final topology for all the mappings*

$$Spp(R) \to \mathcal{A} \; ; \; J \longmapsto [a] \in A/_{\hat{J}}$$

for any $a \in A$. *The mapping*

$$p : \mathcal{A} \to Spp(R) \; ; \; [a] \in A/_{\hat{J}} \mapsto J$$

is a local homeomorphism and A *is isomorphic to the module of continuous sections of* p.

For any pure ideal I and any element $a \in A$ the composite

$$\mathcal{O}_I \hookrightarrow Spp(R) \to \mathcal{A} \; ; \; J \longmapsto [a] \in A/_{\hat{J}}$$

is continuous by definition of the topology on $\mathcal{A}$. Now consider some element $x \in \nabla A(I)$; there is a pure covering $I = \underset{k\in K}{+} I_k$ such that for any k,

$$x\big|_{I_k} = [a_k] \in A/_{\complement\, AI_k}.$$

Consider the mapping

$$\mathcal{O}_I \to \mathcal{A} \; ; \; J \mapsto [x] \in A/_{\hat{J}}.$$

For any k, the restriction of this mapping to $\mathcal{O}_{I_k}$ is

$$\mathcal{O}_{I_k} \hookrightarrow \mathcal{O}_I \to \mathcal{A} \; ; \; J \mapsto [x] = [a_k] \in A/_{\hat{J}},$$

and we have already proved the continuity of this mapping. But the $\mathcal{O}_{I_k}$'s form an open covering of $\mathcal{O}_I$ (theorem 32); thus the mapping $\mathcal{O}_I \to \mathcal{R}$ is continuous.

Finally the topology on $\mathcal{R}$ is also the final topology for all the mappings

$$\mathcal{O}_I \to \mathcal{A} \; ; \; J \mapsto [x] \in A/_{\hat{J}}$$

for any pure ideal I and any $x \in \nabla A(I)$. Therefore theorem 53 is just the "espace étalé" version of theorem 49 (cfr. [9]). The module structure on the set of continuous sections of p is defined pointwise from the module structure of each stalk in $A/\hat{J}$. ■

§ 7. A COUNTEREXAMPLE FOR PURE SHEAF REPRESENTATIONS

In § 6, we have described a sheaf representation ∇R of a ring R. In terms of "espace étalé", the stalk at a purely prime ideal J is $R/\tilde{J}$ (proposition 46).

It is natural to ask whether there exists some sheaf representation of R on its pure spectrum whose stalk at J is just R/J (Pierce's representation is of this kind). We will show that this cannot be true in general.

Consider a ring R with a single non trivial pure ideal I. For example, the ring of linear endomorphisms of an infinite countable dimensional vector space (example 35) or the ring of triangular 2 × 2 matrices on some field (example 36). For such a ring, the pure spectrum is the Sierpienski space

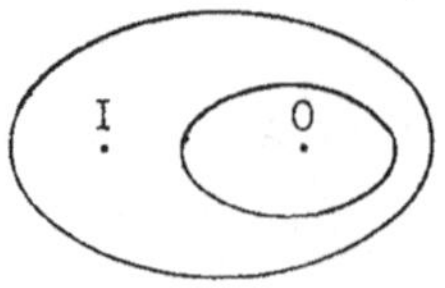

Now, consider a local homeomorphism $p : R \to Spp(R)$ where $p^{-1}(0) = R/0 = R$ and $p^{-1}(I) = R/I$. Take $r \in R$ in the stalk at 0; the local homeomorphism condition implies that r is an open point in R because $p(r) = 0$ is open in Spp(R). Thus the stalk at 0 is discrete.

Consider $[r] \in R/I$. The local homeomorphism condition implies the existence of some $s_{[r]} \in R$ in the stalk at 0 such that the pair $\{[r], s_{[r]}\}$ is open in R. Moreover, for any $[r] \in R/I$, there is a unique s such that $\{[r], s\}$ is open. Indeed if $\{[r], s\}$ and $\{[r], s'\}$ are open with $s \neq s'$, the intersection of these two open subsets is just $[r]$ which is thus an open point. But $p([r]) = I$ is not open in Spp(R) and this contradicts the fact that p is a local homeomorphism. So the uniqueness of $s_{[r]}$ is established and we have in fact a mapping

$$s : R/I \to R \; ; \; [r] \mapsto s_{[r]}.$$

The open subsets described above form a base for the topology of R. Indeed, the intersection of two such open subsets is

$$\{r\} \cap \{s\} = \begin{cases} \{r\} & \text{if } r = s \\ \phi & \text{otherwise.} \end{cases}$$

$$\{[r], s_{[r]}\} \cap \{s\} = \begin{cases} s_{[r]} & \text{if } s_{[r]} = s \\ \phi & \text{otherwise.} \end{cases}$$

$$\{[r], s_{[r]}\} \cap \{[r'], s_{[r']}\} = \begin{cases} \{[r], s_{[r]}\} & \text{if } [r] = [r'] \\ \{s_{[r]}\} & \text{if } [r] \neq [r'] \text{ and } s_{[r]} = s_{[r']} \\ \phi & \text{otherwise.} \end{cases}$$

Clearly, the open subsets of R are just the unions of these basic open subsets. In other words, $U \subseteq R$ is open if

$$[r] \in U \Rightarrow s_{[r]} \in U.$$

We are now able to compute the continuous sections of p. First observe that for any $[r] \in R/I$, the mapping

$$\sigma_{[r]} : \mathrm{Spp}(R) \to R \; ; \; I \mapsto [r] \; ; \; 0 \mapsto s_{[r]}$$

is continuous. Indeed, for any open subset U

$$\sigma_{[r]}^{-1}(U) = \begin{cases} \phi & \text{if } [r] \notin U,\ s_{[r]} \notin U \\ \{0\} & \text{if } [r] \notin U,\ s_{[r]} \in U \\ \{0,I\} & \text{if } [r] \in U \text{ and thus } s_{[r]} \in U. \end{cases}$$

Now consider a continuous section

$$\sigma : \mathrm{Spp}(R) \to R.$$

We know that $\{\sigma(I), s_{\sigma(I)}\}$ is open in R and thus

$$\sigma^{-1}\{\sigma(I), s_{\sigma(I)}\} = \begin{cases} \{0,I\} & \text{if } \sigma(0) = s_{\sigma(I)} \\ \{I\} & \text{if } \sigma(0) \neq s_{\sigma(I)} \end{cases}$$

is open in Spp(R). As I is not an open point in Spp(R), this implies $\sigma(0) = s_{\sigma(I)}$ and thus $\sigma = \sigma_{\sigma(I)}$. So the continuous sections of p are just the $\sigma_{[r]}$ for any $[r]$ in R/I.

Now, if $p : R \to Spp(R)$ is a sheaf representation of R, the set of continuous sections of p must be a ring for pointwise operations. This means exactly that $s : R/_I \to R$ must be a ring homomorphism. Now a continuous section $\sigma_{[r]}$ is exactly determined by [r] and the ring of global sections is just the ring $R/_I$. As $p : R \to Spp(R)$ is a sheaf representation of R, this ring of continuous sections is isomorphic to R. So we conclude that the rings R and $R/_I$ are isomorphic. This conclusion depends only on the fact that Spp(R) is the Sierpinski space, not on the precise form of R.

In the case of the ring R of example 35, R has a single non trivial 2-sided ideal which is precisely I. This implies that $R/_I$ is a simple ring, i.e. a ring with only the two trivial 2-sided ideals. Indeed, if $J \subseteq R/_I$ is a 2-sided ideal its inverse image $q^{-1}(J)$ along the quotient map $q : R \to R/_I$ is a 2-sided ideal in R and thus $J = q(q^{-1}(J))$ is (0) or $R/_I$. This proves that in the specific case of example 35, R is not isomorphic to $R/_I$. So in that case, there cannot be a sheaf representation $p : R \to Spp(R)$ of R with $p^{-1}(I) = R/_I$ and $p^{-1}(0) = R$.

In the case of example 36, the 2×2 triangular matrices on some field K, compute the "espaces étalés" corresponding to the sheaves ΔR and ∇R. The pure ideals not contained in (0) are I and R and the only pure ideal not contained in I is R.

The stalk of ΔR at I is the inductive limit of (R, R), thus it is $(R, R) \cong R$. The stalk of ΔR at (0) is the inductive limit of $(R, R) \to (I, I)$, thus it is (I, I). Recall that I is the ideal of the matrices of the form

$$\begin{pmatrix} 0 & a \\ 0 & b \end{pmatrix}.$$

There are canonical inclusions of rings

$$K \hookrightarrow R \hookrightarrow K^{2 \times 2},$$

where the first inclusion sends $a \in K$ to the diagonal matrix

$$\begin{pmatrix} a & 0 \\ 0 & a \end{pmatrix}.$$

Any R-linear endomorphism of I is thus K-linear. So (I, I) is contained in the ring of K-linear endomorphisms of $I \cong K^2$ which is just $K^{2 \times 2}$. So any R-linear endomorphism of I has the form

$$\begin{pmatrix} 0 & a \\ 0 & b \end{pmatrix} \longmapsto \begin{pmatrix} \alpha & \beta \\ \gamma & \delta \end{pmatrix}\begin{pmatrix} 0 & a \\ 0 & b \end{pmatrix}.$$

But any such mapping is in fact R-linear; indeed, the R-linearity means that

$$\begin{pmatrix} \alpha & \beta \\ \gamma & \delta \end{pmatrix}\left(\begin{pmatrix} 0 & a \\ 0 & b \end{pmatrix}\begin{pmatrix} x & y \\ 0 & z \end{pmatrix}\right) = \left(\begin{pmatrix} \alpha & \beta \\ \gamma & \delta \end{pmatrix}\begin{pmatrix} 0 & a \\ 0 & b \end{pmatrix}\right)\begin{pmatrix} x & y \\ 0 & z \end{pmatrix}$$

which is obvious. Thus (I, I) is just the ring $K^{2 \times 2}$ and this is the stalk of ΔR at (0).

The peculiar form of the space Spp(R) implies that any covering of a non-empty open subset must contain this open subset. Therefore, the two conditions defining a sheaf vanish in the case of a covering of a non-empty open subset. So the sheaf condition reduces to the condition on the empty open subset : the separation condition means that for a sheaf F, there is at most one element in $F(\phi)$ and the glueing condition means that $F(\phi)$ has at least one element. Finally a sheaf F on Spp(R) is just a presheaf F such that $F(\phi)$ is a singleton.

All this implies that the presheaf

$$\mathcal{O}_J \longmapsto R/_{\complement J} \ ; \ J \text{ pure in } R$$

is already a sheaf and thus equal to ∇R. Clearly $\complement R = (0)$. On the other hand, $\complement I$ is a 2-sided ideal (proposition 16) whose intersection with I is zero. If $\complement I$ contains a non-zero matrix A, the condition $A \notin I$ implies that $a_{11} \neq 0$. But in this case

$$\begin{pmatrix} 1 & 0 \\ 0 & 0 \end{pmatrix} A \begin{pmatrix} 0 & 1 \\ 0 & 0 \end{pmatrix} = \begin{pmatrix} 0 & a_{11} \\ 0 & 0 \end{pmatrix} \in \complement I$$

which is a contradiction since this last matrix is also in I. Finally $\complement I = (0)$.

The stalk of ∇R at I is the inductive limit of $R/_{(0)}$, thus it is R. The stalk of ∇R at (0) in the inductive limit of $R/_0 \to R/_0$, thus it is also R.

Finally, when we compose both "espaces étalés", we conclude that the stalks at (0) are different. For ΔR, it is $K^{2 \times 2}$ and for ∇R, it is just the subring R of $K^{2 \times 2}$.

§ 8. PURE IDEALS IN PRODUCTS OF RINGS

So far in this chapter, the ring R was fixed. In these last two paragraphs, we let R vary and investigate what happens to the pure ideals and the pure spectra.

In this paragraph, we consider the case of a product $\underset{k \in K}{\times} R_k$ of rings. Any product of pure ideals is pure but the converse is generally not true; however it is true when K is finite. Again when K is finite, we are able to describe the purely prime ideals of $\underset{k \in K}{\times} R_k$ and we conclude that the pure spectrum of $\underset{k \in K}{\times} R_k$ is just the disjoint union of the spectra of the rings R_k.

Proposition 54.

Let $(R_k)_{k \in K}$ be a family of rings and for any $k \in K$, I_k a pure ideal in R_k. In this case, $\underset{k \in K}{\times} I_k$ is a pure ideal of the ring $\underset{k \in K}{\times} R_k$.

Clearly, $\underset{k \in K}{\times} I_k$ is a 2-sided ideal. Let $(i_k)_{k \in K}$ be some element in this ideal. Then for any k, choose $\varepsilon_k \in I_k$ such that $i_k \, \varepsilon_k = i_k$. The equality

$$(i_k)_{k \in K} \cdot (\varepsilon_k)_{k \in K} = (i_k)_{k \in K}$$

holds in $\underset{k \in K}{\times} I_k$. ∎

A pure ideal I in a product $\underset{k \in K}{\times} R_k$ of rings is generally not a product of pure ideals I_k in the R_k's. For example, if K is an infinite set, take I to be the ideal of those families $(r_k)_{k \in K}$ such that all but a finite number of its elements r_k are zero. I is clearly 2-sided, left and right pure : the unit $(\varepsilon_k)_{k \in K}$ of $(r_k)_{k \in K}$ can be choosen to be 1 when $r_k \neq 0$ and 0 otherwise. I is not a product of pure ideals in the R_k's as soon as infinitely many of the R_k's are not the zero ring. But if K is finite, any pure ideal in $\underset{k \in K}{\times} R_k$ is of the form $\underset{k \in K}{\times} I_k$ with I_k pure in R_k. To prove this, we need the following lemma.

Lemma 55.

Let $f : R \to S$ be a surjective homomorphism of rings. For any pure ideal I in R, f(I) is a pure ideal of S.

Clearly f(I) is a subgroup of S. Now for any $s \in S$ and $f(i) \in f(I)$, we can write $s = f(r)$ and therefore

$$s \,.\, f(i) = f(r) \,.\, f(i) = f(r\, i) \in f(I).$$
$$f(i) \,.\, s = f(i) \,.\, f(r) = f(i\, r) \in f(I)$$

and f(I) is a 2-sided ideal. Moreover if $f(i) \in f(I)$, choose $\varepsilon \in I$ such that $i\,\varepsilon = i$. We have $f(\varepsilon) \in f(I)$ and

$$f(i) \,.\, f(\varepsilon) = f(i\,\varepsilon) = f(i);$$

so f(I) is pure. ■

Proposition 56.

Let $(R_k)_{k\in K}$ be a finite family of rings and I a pure ideal in $\underset{k\in K}{\times} R_k$. For any $k \in K$ there is a pure ideal I_k of R_k such that $I = \underset{k\in K}{\times} I_k$.

For any $k \in K$, consider the canonical projection

$$p_k : \underset{k\in K}{\times} R_k \to R_k.$$

By lemma 55, each $I_k = p_k(I)$ is a pure ideal in R_k. We shall prove that $I = \underset{k\in K}{\times} I_k$. Obviously we have the inclusion $I \subseteq \underset{k\in K}{\times} I_k$.

Now consider $(i_k)_{k\in K}$ an element in $\underset{k\in K}{\times} I_k$. For any k, i_k is in $p_k(I)$; thus there is some element $(i_k^\ell)_{\ell\in K}$ in I with $i_k^k = i_k$. Choose $(\varepsilon_k)_{k\in K}$ in I, a unit for all $(i_k^\ell)_{\ell\in K}$. Thus for any k and any ℓ, we have

$$i_k^\ell \,.\, \varepsilon_\ell = i_k^\ell.$$

In particular, for any k

$$i_k \,.\, \varepsilon_k = i_k^k \,.\, \varepsilon_k = i_k^k = i_k.$$

So we have the equality

$$(i_k)_{k\in K} \,.\, (\varepsilon_k)_{k\in K} = (i_k)_{k\in I}$$

with $(\varepsilon_k)_{k\in K}$ in I. This implies that $(i_k)_{k\in K}$ is also an element in I. ■

From proposition 56, it follows easily that the lattice of pure ideals of $\underset{k\in K}{\times} R_k$ is isomorphic to the product of the lattices of pure ideals of the

R_k's. Taking the associated Stone spaces, we deduce that the pure spectrum of $\underset{k\in K}{\times} R_k$ is the disjoint union of the pure spectra of the R_k's. We propose a more direct proof of this fact. This proof requires some lemmas.

Lemma 57.

Let $f : R \to S$ *be a surjective homomorphism of rings and* I, J *two pure ideals of* R. *Then,* $f(I \cap J) = f(I) \cap f(J)$.

Clearly $f(I \cap J) \subseteq f(I) \cap f(J)$. Now if $f(i) = f(j)$ with $i \in I$, $j \in J$, choose $\varepsilon \in I$ such that $i\,\varepsilon = i$

$$f(i) = f(i\,\varepsilon) = f(i)\,f(\varepsilon) = f(j)\,f(\varepsilon) = f(j\,\varepsilon)$$

and $j\,\varepsilon \in J\,I = J \cap I$ (proposition 2). Thus $f(I) \cap f(J) \subseteq f(I \cap J)$. ■

Lemma 58.

Let $f : R \to S$ *be a surjective homomorphism of rings and* J *a purely prime ideal in* S. *The pure part* $\overset{\circ}{\widehat{f^{-1}(J)}}$ *of* $f^{-1}(J)$ *is a purely prime ideal in* R.

$J = f\,f^{-1}(J)$ is proper, thus $\overset{\circ}{\widehat{f^{-1}(J)}}$ is a proper ideal. Take I_1, I_2 pure in R such that $I_1 \cap I_2 \subseteq \overset{\circ}{\widehat{f^{-1}(J)}}$. This implies, by lemma 57, that

$$\begin{aligned} f(I_1) \cap f(I_2) &= f(I_1 \cap I_2) \\ &\subseteq f\,\overset{\circ}{\widehat{f^{-1}(J)}} \\ &\subseteq f\,f^{-1}(J) = J. \end{aligned}$$

By lemma 55, $f(I_1)$ and $f(I_2)$ are pure and since J is purely prime,

$$f(I_1) \subseteq J \text{ or } f(I_2) \subseteq J.$$

This implies immediately that

$$I_1 \subseteq f^{-1}(J) \text{ or } I_2 \subseteq f^{-1}(J)$$

and since I_1, I_2 are pure

$$I_1 \subseteq \overset{\circ}{\widehat{f^{-1}(J)}} \text{ or } I_2 \subseteq \overset{\circ}{\widehat{f^{-1}(J)}}.$$ ■

Proposition 59.

Let $(R_k)_{k\in K}$ *be a finite family of rings. The pure spectrum of* $\underset{k\in K}{\times} R_k$ *is the disjoint union of the pure spectra of the rings* R_k.

For any purely prime ideal J_ℓ in R_ℓ, we obviously have

$$p_\ell^{-1}(J_\ell) = \underset{k\in K}{\times} J_k \text{ where } J_k = R_k \text{ for } k \neq \ell.$$

As $p_\ell^{-1}(J_\ell)$ is not the whole space $\underset{k\in K}{\times} R_k$ and as it is pure (proposition 54), we deduce by lemma 58 that $p_\ell^{-1}(J_\ell)$ is purely prime in $\underset{k\in K}{\times} R_k$. We will prove that these ideals $p_\ell^{-1}(J_\ell)$ are the only purely prime ideals of $\underset{k\in K}{\times} R_k$.

Consider J purely prime in $\underset{k\in K}{\times} R_k$. By proposition 56, $J = \underset{k\in K}{\times} J_k$ with J_k pure in R_k. As J is proper, at least one J_k is a proper ideal in R_k. In fact, exactly one J_k is a proper ideal in R_k. Indeed, if J_ℓ and J_m were proper respectively in R_ℓ and R_m with $\ell \neq m$, consider the two pure ideals

$$I^\ell = \underset{k\in I}{\times} I_k^\ell \text{ where } \begin{cases} I_\ell^\ell = R_\ell \\ I_k^\ell = J_k \text{ if } k \neq \ell. \end{cases}$$

$$I^m = \underset{k\in I}{\times} I_k^m \text{ where } \begin{cases} I_m^m = R_m \\ I_k^m = J_k \text{ if } k \neq m. \end{cases}$$

From their definition, it follows immediately that $I^\ell \cap I^m = J$ and $I^\ell \nsubseteq J$, $I^m \nsubseteq J$. Thus J is not purely prime.

Finally the purely prime ideals of $\underset{k\in K}{\times} R_k$ are exactly the $p_\ell^{-1}(J_\ell)$ where J_ℓ is purely prime in R_ℓ. So the assignment

$$J_\ell \mapsto p_\ell^{-1}(J_\ell)$$

describes a bijection between the disjoint union of the $\mathrm{Spp}(R_k)$'s and $\mathrm{Spp}(\underset{k\in K}{\times} R_k)$. We need to show that this is an homeomorphism.

An open subset in $\underset{k\in K}{\times} R_k$ is just a pure ideal $I = \underset{k\in K}{\times} I_k$ (proposition 56). This open subset contains the point $J = \underset{k\in K}{\times} J_k \in \mathrm{Spp}(\underset{k\in K}{\times} R_k)$, where J_ℓ is purely prime in R_ℓ, if and only if $I \nsubseteq J$. But for $k \neq \ell$, $I_k \subseteq R_k = J_k$; thus $I_\ell \nsubseteq J_\ell$. Finally, an open neighbourhood of $J = p_\ell^{-1}(J_\ell)$ in $\mathrm{Spp}(\underset{k\in K}{\times} R_k)$ is just a family $(I_k)_{k\in K}$ of open subsets in each $\mathrm{Spp}(R_k)$ in such a way that the point J_ℓ belongs to the open subset I_ℓ. But this is exactly the disjoint union of the topological spaces $\mathrm{Spp}(R_k)$. ∎

§ 9. CHANGE OF BASE RING

In this last paragraph, we study the action of a ring homomorphism $f : R \to S$ on pure ideals and the pure spectra.

This problem is not easy at all and is still open in the general case. The difficulty comes from the fact that the image of an ideal is generally not an ideal and the inverse image of a pure ideal is generally not a pure ideal. For example, consider the ring inclusion $\mathbb{Z} \hookrightarrow \mathbb{Q}$; $\mathbb{Z}$ is an ideal in $\mathbb{Z}$ but not in $\mathbb{Q}$. On the other hand, consider $\mathbb{Z} \to \mathbb{Z}/_{n\mathbb{Z}}$; (0) is a pure ideal in $\mathbb{Z}/_{n\mathbb{Z}}$ but its inverse image $n\mathbb{Z}$ is not a pure ideal in $\mathbb{Z}$ as soon as $n \neq 0$, $n \neq 1$.

To overcome these difficulties, it seems reasonable to use natural constructions like "ideal generated by a subset" or "pure part of an ideal". This approach of the problem produces a continuous mapping $\mathrm{Spp}(f) : \mathrm{Spp}(S) \to \mathrm{Spp}(R)$ in two particular cases : when f is surjective or when S is commutative. Moreover, if $R \xrightarrow{f} S \xrightarrow{g} T$ is a composite of two ring homomorphisms such that each of them is either surjective or with codomain commutative, then $\mathrm{Spp}(f) \circ \mathrm{Spp}(g) = \mathrm{Spp}(g \circ f)$.

In § 8, we described how pure ideals and purely prime ideals can be transformed by a surjective ring homomorphism (lemmas 55, 57, 58). We start with analogous lemmas in the case of a ring homomorphism $f : R \to S$ with S commutative.

Lemma 60.

Let $f : R \to S$ *be a ring homomorphism with* S *commutative. For any pure ideal* I *in* R, *the ideal* $f(I) . S$ *generated by* $f(I)$ *in* S *is pure.*

Take $i \in I$ and $s \in S$, thus $f(i) . s \in f(I) . S$. Choose ε in I such that $i\,\varepsilon = i$. We have

$$f(i) . s = f(i\,\varepsilon) . s = f(i) . f(\varepsilon) . s = f(i) . s . f(\varepsilon)$$

and $f(\varepsilon) \in f(I) . S$. Thus $f(I) . S$ is pure. ∎

<u>Lemma 61.</u>

Let $f : R \to S$ *be a ring homomorphism with* S *commutative. Let* I, J *be two pure ideals in* R. *Then*

$$f(I \cap J) \,.\, S = (f(I) \,.\, S) \cap (f(J) \,.\, S).$$

Clearly $f(I \cap J) \,.\, S \subseteq (f(I) \,.\, S) \cap (f(J) \,.\, S)$. Now consider

$$f(i) \,.\, s = f(j) \,.\, s' \;;\; i \in I \;;\; j \in J \;;\; s, s' \in S$$

some element in $(f(I) \,.\, S) \cap (f(J) \,.\, S)$. Choose $\varepsilon \in I$ such that $i\,\varepsilon = i$.

$$\begin{aligned} f(i) \,.\, s &= f(i\,\varepsilon) \,.\, s \\ &= f(i) \,.\, f(\varepsilon) \,.\, s \\ &= f(i) \,.\, s \,.\, f(\varepsilon) \\ &= f(j) \,.\, s' \,.\, f(\varepsilon) \\ &= f(j) \,.\, f(\varepsilon) \,.\, s' \\ &= f(j\,\varepsilon) \,.\, s' \\ &\in f(I \cap J) \,.\, S \qquad \text{(proposition 2).} \end{aligned}$$

Thus the equality holds. ■

<u>Lemma 62.</u>

Let $f : R \to S$ *be a ring homomorphism with* S *commutative. Let* J *be a purely prime ideal in* S. *The pure part* $\overset{\circ}{\widehat{f^{-1}(J)}}$ *of* $f^{-1}(J)$ *is a purely prime ideal in* R.

Take I_1, I_2 pure in R such that $I_1 \cap I_2 \subseteq \overset{\circ}{\widehat{f^{-1}(J)}}$. This implies by lemma 61 :

$$\begin{aligned} (f(I_1) \,.\, S) \cap (f(I_2) \,.\, S) &= f(I_1 \cap I_2) \,.\, S \\ &\subseteq f(\overset{\circ}{\widehat{f^{-1}(J)}}) \,.\, S \\ &\subseteq f(f^{-1}(J)) \,.\, S \\ &\subseteq J \,.\, S \\ &\subseteq J. \end{aligned}$$

By lemma 60, $f(I_1) \,.\, S$ and $f(I_2) \,.\, S$ are pure and since J is purely prime

$$f(I_1) \,.\, S \subseteq J \quad \text{or} \quad f(I_2) \,.\, S \subseteq J.$$

This implies

$$f(I_1) \subseteq J \quad \text{or} \quad f(I_2) \subseteq J$$
$$f^{-1}\, f(I_1) \subseteq f^{-1}(J) \quad \text{or} \quad f^{-1}\, f(I_2) \subseteq f^{-1}(J)$$
$$I_1 \subseteq f^{-1}(J) \quad \text{or} \quad I_2 \subseteq f^{-1}(J),$$

and finally, since I_1 and I_2 are pure

$$I_1 \subseteq \overset{\circ}{\widehat{f^{-1}(J)}} \quad \text{or} \quad I_2 \subseteq \overset{\circ}{\widehat{f^{-1}(J)}}.$$

We still need to prove that $\overset{\circ}{\widehat{f^{-1}(J)}}$ is a proper ideal of R. But if $\overset{\circ}{\widehat{f^{-1}(J)}} = R$, then $f^{-1}(J) = R$ and thus $J \supseteq f(R)$. So $1 = f(1) \in J$ and J is not proper, which is a contradiction. ∎

<u>Proposition 63.</u>

Let $f : R \to S$ *be a ring homomorphism which is either surjective or with codomain commutative. Define a mapping*

$$\mathrm{Spp}(f) : \mathrm{Spp}(S) \to \mathrm{Spp}(R);\ J \longmapsto \overset{\circ}{\widehat{f^{-1}(J)}}.$$

This mapping is continuous.

The mapping Spp(f) is well defined by lemmas 58 and 62. To prove the continuity, choose $\mathcal{O}_I$ an open subset of Spp(R), i.e. a pure ideal I of R. The points of $\mathcal{O}_I$ are the purely prime ideals J' of R such that $I \not\subseteq J'$. We must prove that $\mathrm{Spp}(f)^{-1}(\mathcal{O}_I)$ is open in Spp(S). In fact we shall prove that

$$\mathrm{Spp}(f)^{-1}(\mathcal{O}_I) = \mathcal{O}_{\langle f(I)\rangle}$$

where $\langle f(I)\rangle$ is the 2-sided ideal generated by f(I) in S; $\langle f(I)\rangle$ is pure by lemmas 55 and 60.

Consider a purely prime ideal J of S.

$$\begin{aligned} J \in \mathrm{Spp}(f)^{-1}(\mathcal{O}_I) &\iff \mathrm{Spp}(f)(J) \in \mathcal{O}_I \\ &\iff I \not\subseteq \overset{\circ}{\widehat{f^{-1}(J)}} \\ &\iff I \not\subseteq f^{-1}(J) \qquad \text{(because I is pure)} \\ &\iff fI \not\subseteq J \\ &\iff \langle fI\rangle \not\subseteq J \\ &\iff J \in \mathcal{O}_{\langle f(I)\rangle} \end{aligned}$$

This proves the continuity of Spp(f). ■

Proposition 64.

Let $f : R \to S$ *and* $g : S \to T$ *be two ring homomorphisms such that each of them is either surjective or with codomain commutative. Then* $g \circ f$ *is either surjective or with codomain commutative and*

$$\mathrm{Spp}(g \circ f) = \mathrm{Spp}(f) \circ \mathrm{Spp}(g).$$

Suppose T is not commutative; then g is surjective. But then T is a quotient ring of S; thus S cannot be commutative. This implies that f is surjective and finally $g \circ f$ is surjective. So $\mathrm{Spp}(g \circ f)$ is defined.

Take J a purely prime ideal in T. We have

$$\mathrm{Spp}(g)(J) = \overset{\circ}{\widehat{g^{-1}(J)}}$$

$$\mathrm{Spp}(f) \circ \mathrm{Spp}(g)(J) = \overset{\circ}{\widehat{f^{-1}(\overset{\circ}{\widehat{g^{-1}(J)}})}}$$

$$\mathrm{Spp}(g \circ f)(J) = \overset{\circ}{\widehat{f^{-1}\, g^{-1}(J)}}.$$

But $\overset{\circ}{\widehat{f^{-1}(\overset{\circ}{\widehat{g^{-1}(J)}})}}$ is a pure ideal contained in $f^{-1}\, g^{-1}(J)$, thus the following inclusion holds

$$\overset{\circ}{\widehat{f^{-1}(\overset{\circ}{\widehat{g^{-1}(J)}})}} \subseteq \overset{\circ}{\widehat{f^{-1}\, g^{-1}(J)}}.$$

To prove the converse inclusion, consider the following diagrams in R and S

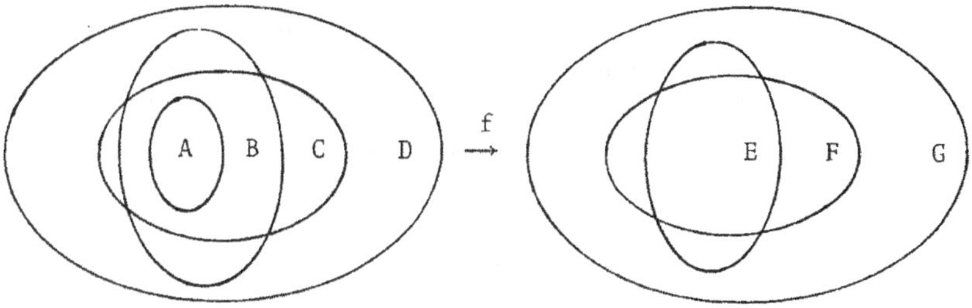

$$A = \overset{\circ}{\widehat{f^{-1}(\overset{\circ}{\widehat{g^{-1}(J)}})}} \qquad E = \overset{\circ}{\widehat{g^{-1}(J)}}$$

$$B = f^{-1}(\overset{\circ}{\widehat{g^{-1}(J)}}) \qquad F = <f(\overset{\circ}{\widehat{f^{-1}\ g^{-1}(J)}})>$$

$$C = \overset{\circ}{\widehat{f^{-1}\ g^{-1}(J)}} \qquad G = g^{-1}(J).$$

$$D = f^{-1}\ g^{-1}(J)$$

Again the brackets notation < > denotes the generated 2-sided ideal. A, B, C, D, E, F, G are 2 sided ideals; A, C, E, F are pure ideals and A, C, E are purely prime ideals (lemmas 55, 58 and 60). E is the pure part of G and F is pure, $F \subseteq G$; this implies $F \subseteq E$ and thus $f^{-1}(F) \subseteq f^{-1}(E) = B$. Moreover

$$C = \overset{\circ}{\widehat{f^{-1}\ g^{-1}(J)}} \subseteq f^{-1}\ f(\overset{\circ}{\widehat{f^{-1}\ g^{-1}(J)}}) \subseteq f^{-1}(F) = B.$$

But C is pure, $C \supseteq A$ and A is the pure part of B, thus $A = C$. This is just what we needed to prove. ■

Counterexample 65.

We conclude this paragraph with an example of a ring homomorphism $f : R \to S$ which does not produce a continuous mapping $Spp(S) \to Spp(R)$ when we apply to it the constructions of proposition 63.

Consider a field K, the commutative ring $R = K^2$ and the non-commutative ring S of triangular 2×2 matrices on K. Take f to be the inclusion

$$f : R \to S \quad ; \quad (a, b) \longmapsto \begin{pmatrix} a & 0 \\ 0 & b \end{pmatrix}.$$

Clearly f is not surjective and S is not commutative. The pure ideals of R are (o), $(o) \times K$, $K \times (o)$, $K \times K$ because $a\,\varepsilon = a$ in K implies $a = o$ or $\varepsilon = 1$, since K is a field. In particular, the purely prime ideals of R are $(o) \times K$ and $K \times (o)$. On the other hand (example 36) the pure ideals of S are S, the ideal J of matrices with first column zero and (o). We have

$$\overset{\circ}{\widehat{f^{-1}(J)}} = f^{-1}(J) = (o) \times K$$

$$\overset{\circ}{\widehat{f^{-1}(o)}} = f^{-1}(o) = (o).$$

$(o) \times K$ is purely prime in R but (o) is not purely prime in R.

This example also yields a situation where the pure part construction does not commute with the inverse image. Let I be the 2-sided (and left pure) ideal of S of those matrices with second row zero (see example 36). The pure part of I is just (o) and thus $f^{-1}(\overset{\circ}{I}) = (o)$. On the other hand, $f^{-1}(I) = K \times (o)$ which is pure in S.

Moreover, consider the ring homomorphism

$$g : S \to R \quad ; \quad \begin{pmatrix} a & c \\ a & b \end{pmatrix} \longmapsto (a,b)$$

which is surjective and has a commutative image. The composite $g \circ f$ is just the identity and therefore it induces the identity mapping on Spp(R); in particular $\mathrm{Spp}(g \circ f)(K \times (o)) = K \times (o)$. On the other hand, $g^{-1}(K \times (o)) = I$ and the pure part of I is (o); thus $\mathrm{Spp}(g)(K \times (o)) = (o)$. If we apply the construction of proposition 63 to $\mathrm{Spp}(g)(K \times (o))$, we find (o) in R, which is not purely prime and which is not equal to $\mathrm{Spp}(g \circ f)(K \times (o))$.

CHAPTER 8 : GELFAND RINGS

This last chapter develops the results of chapter 7 in the special context of Gelfand rings. A detailed study of the structure of maximal and pure ideals in Gelfand rings allows us to generalize to arbitrary Gelfand rings the results of Bkouche (cfr. [3]) on commutative Gelfand rings. As a consequence, our theory reduces to that of Pierce in the case of von Neumann regular rings (cfr. [19]).

A ring is right Gelfand if its right maximal ideals satisfy a separation condition (cfr. [3], [16]). Mulvey has proved (cfr. [16]) that right Gelfand is equivalent to left Gelfand. Here we explain the reason for this : in a Gelfand ring, the left maximal ideals are exactly the right maximal ideals (§§ 1 - 3).

But a Gelfand ring can equivalently be characterized by properties of its pure ideals (§ 3). The properties of pure ideals in a Gelfand ring are deduced from an interesting formula describing the pure part of an ideal I (§ 2) :

$$\overset{\circ}{I} = \{a \in R \mid \exists \, \varepsilon \in I \quad a\,\varepsilon = a\}.$$

From this formula, we deduce characterizations of Gelfand rings in terms of pure ideals (theorem 31) and we deduce also the important fact that in a Gelfand ring, the left pure ideals are exactly the right pure ideals.

This coincidence between left and right pure ideals implies that for a Gelfand ring, the left pure spectrum is exactly the right pure spectrum; moreover this pure spectrum is compact Hausdorff and homeomorphic to the usual maximal spectrum (§ 4). Moreover, the sheaf representations ∇R and ΔR coincide in the case of Gelfand rings and their stalks are local rings (§ 5). This extends a result of Bkouche for commutative Gelfand rings (cfr. [3]). The properties of § VII - 9 on the change of base ring also extend to Gelfand rings (§ 6).

The last paragraph is devoted to examples of Gelfand rings. Local rings and von Neumann regular rings are such examples. A very characteristic example of a Gelfand ring which is generally not regular is given by the ring $C(X, \mathbb{R})$ of real valued functions on some topological space X. From this example, using rings of matrices or quaternion rings, we construct some examples of non commutative Gelfand rings (§ 7).

Several results of §§ 1 - 2 - 3 were obtained first for commutative Gelfand rings. Their translation into the non commutative case was made possible by a fruitful collaboration with Harold Simmons. Several other interesting results on Gelfand rings and pure representations will be found in further papers by him.

§ 1. GELFAND RINGS

We define the notion of right-Gelfand ring and we will show that it is equivalent to the notion of left Gelfand ring. This result was known by Mulvey (cfr. [16]) from an abstract categorical argument; here we make the reason of this fact very explicit : we show that in a Gelfand ring, an ideal is right maximal if and only if it is left maximal. We start with some well-known easy lemmas on maximal ideals. R is an arbitrary ring with unit; when nothing is specified, all ideals are right ideals.

Lemma 1.

Let M *be a maximal ideal and* $M^{\wedge}$ *the greatest 2-sided ideal contained in* M. *Then*

$$M^{\wedge} = \bigcap_{r \in R} [M : r]$$

$$[M : r] = \{x \in R \mid r x \in M\}.$$

$$M^{\wedge} = \{x \in R \mid \forall r \in R \quad r x \in M\}.$$

$M^{\wedge}$ is obviously a 2-sided ideal; choosing $r = 1$, we deduce $M^{\wedge} \subseteq M$. Let I be a 2-sided ideal, $I \subseteq M$ and $x \in I$; for any $r \in R$, $r x \in I \subseteq M$ thus $x \in M^{\wedge}$; so $I \subseteq M^{\wedge}$ and $M^{\wedge}$ is the greatest 2-sided ideal in M. ■

Lemma 2.

Let M *be a maximal ideal. Then* $M^{\wedge}$ *is a prime ideal.*

By lemma 1, $M^{\wedge} = \bigcap_{r \in R} [M : r]$. Now take $a, b \in R$ such that $a R b \subseteq M^{\wedge}$. Then

$$a \notin M^{\wedge} \Rightarrow \exists r \in R \quad r a \notin M.$$

Therefore $r a R + M = R$ and

$$1 = r\,a\,s + m \quad ; \quad s \in R \quad ; \quad m \in M.$$

$R\,b = r\,a\,s\,R\,b + m\,R\,b \subseteq M.$
By lemma 1, we deduce $b \in M^\wedge$. ■

Lemma 3.

Let M *be a maximal ideal and* $r \notin M$. *Then* $[M : r]$ *is a maximal ideal.*

$[M : r]$ is proper because $1 \in [M : r]$ implies $r \in M$. Take $s \notin [M : r]$, thus $r\,s \notin M$. We have $r\,s\,R + M = R$, thus

$$r = r\,s\,t + m \quad ; \quad t \in R \quad ; \quad m \in M;$$
$$r(1 - s\,t) = m \in M.$$

This implies $1 - s\,t \in [M : r]$ and thus

$$1 \in s\,t + [M : r] \subseteq s\,R + [M : r].$$

Thus if we adjoin to $[M : r]$ any element $s \notin [M : r]$, we generate the whole ring. This says exactly that $[M : r]$ is a maximal ideal. ■

Definition 4.

A ring R *is a right Gelfand ring if for any distinct maximal ideals* $M \neq N$, *there exist* $r \in R$ *and* $s \in R$ *such that* $r \notin M$, $s \notin N$ *and* $r\,R\,s = 0$.

Proposition 5.

In a Gelfand ring, a 2-sided prime ideal is contained in a unique maximal ideal.

Consider P a 2-sided prime ideal and M, N two maximal ideals, $P \subseteq M$, $P \subseteq N$, $M \neq N$. Choose

$$r \notin M \qquad s \notin N \qquad r\,R\,s = 0 \subseteq P \text{ (definition 4).}$$

Since P is prime, $r \in P$ or $s \in P$ and thus $r \in M \cap N$ or $s \in M \cap N$, which is a contradiction. ■

Proposition 6.

Let M *be a maximal ideal in a Gelfand ring and* $r \notin M$. *Then* $M = [M : r]$.

We know that $M^\wedge = \bigcap_{s \in R} [M : s]$ (lemma 1); thus $M^\wedge \subseteq [M : r]$ and $[M : r]$ is maximal (lemma 3). But $M^\wedge$ is prime (lemma 2) and $M^\wedge \subseteq M$, thus $M = [M : r]$ (proposition 5). ■

Proposition 7.

In a Gelfand ring, any maximal ideal is 2-sided.

Let M be a maximal ideal and $r \in R$. If $r \in M$ and $m \in M$, $r\,m \in M$. If $r \notin M$ and $m \in M$, $M = [M : r]$ (proposition 6) and thus

$$m \in M \Rightarrow m \in [M : r] \Rightarrow r\,m \in M. \qquad \blacksquare$$

Proposition 8.

In a Gelfand ring, a maximal ideal is completely prime.

Suppose M is maximal and $r\,s \in M$; we must prove $r \in M$ or $s \in M$. If $r \notin M$, by proposition 6, $M = [M : r]$ and $s \in [M : r]$ because $r\,s = o \in M$. ■

Proposition 9.

In a right Gelfand ring, any right maximal ideal is also left maximal.

Let M be a right maximal ideal; M is also a left ideal (proposition 7). Consider a left ideal I such that

$$M \underset{\neq}{\subset} I.$$

Choose $r \in I \setminus M$; thus $r \notin M$ and $r\,R + M = R$. Consider

$$1 = r\,s + m \quad ; \quad s \in R \quad ; \quad m \in M.$$

We compute

$$(1 - s\,r)s = s - s\,r\,s = s(1 - r\,s) = s\,m \in M,$$

since M is 2-sided. But $s \notin M$ because $s \in M$ implies $1 = r\,s + m \in M$. By proposition 8, $1 - s\,r \in M \subseteq I$. But $s\,r \in I$ since $r \in I$ and I is left sided. So $1 \in I$ and $I = R$. This proves that M is a maximal left ideal. ■

In order to conclude that in a Gelfand ring, right maximal ideals coincide with left maximal ideals, we still need to prove that in a Gelfand ring, any left maximal ideal is 2-sided. This will be deduced from the study of pure ideals in Gelfand rings (§ 2).

§ 2. PURE PART OF AN IDEAL IN A GELFAND RING

This paragraph is devoted to the computation of the pure part of an ideal in a Gelfand ring. This pure part will be the set of all those elements of the ring which have a unit in the ideal. The proofs are generally given first for maximal ideals and then generalized to arbitrary ideals. Again, when nothing is specified, everything is specified on the right. We start with a definition.

Definition 10.

Let I *be an ideal in a ring* R.

The "unit part" of I *is the set*

$$\varepsilon(I) = \{a \in R \mid \forall\, r \in R \quad \exists\, \varepsilon \in I \quad a\,r = a\,r\,\varepsilon\}.$$

Proposition 11.

The "unit part" of an ideal is a 2-sided ideal.

$\varepsilon(I)$ is obviously stable by multiplication on the left and on the right. We must prove that it is also stable by addition. Take a, b in $\varepsilon(I)$ and $r \in R$. Choose ε in I such that $a\,r = a\,r\,\varepsilon$. Consider $b \in \varepsilon(I)$ and $r(1 - \varepsilon) \in R$; choose $\varphi \in I$ such that

$$b\,r(1 - \varepsilon)\varphi = b\,r(1 - \varepsilon).$$

Finally, $\varepsilon + \varphi - \varepsilon\,\varphi \in I$ and

$$\begin{aligned}(a + b)r\,(\varepsilon + \varphi - \varepsilon\,\varphi) &= a\,r\,\varepsilon + a\,r\,\varphi - a\,r\,\varepsilon\,\varphi + b\,r\,\varepsilon + b\,r\,\varphi - b\,r\,\varepsilon\,\varphi \\ &= a\,r + a\,r\,\varphi - a\,r\,\varphi + b\,r\,\varepsilon + b\,r(1 - \varepsilon)\varphi \\ &= a\,r + b\,r\,\varepsilon + b\,r(1 - \varepsilon) \\ &= a\,r + b\,r \\ &= (a + b)r.\end{aligned}$$

■

Proposition 12.

For a maximal ideal M, $\varepsilon(M)$ *is contained in* M.

Take $a \in \varepsilon(M)$ and $r = 1 \in R$; choose $\varepsilon \in M$ such that $a = a\,\varepsilon$. Because M is 2-sided (proposition 7) and $\varepsilon \in M$, we deduce $a \in M$. ■

The proof of proposition 12 does not work for an arbitrary ideal I (which is not left-sided). However, the result is still valid for an arbitrary ideal, but to prove it, we need some more lemmas.

Lemma 13.

Let R *be a Gelfand ring,* M *a maximal ideal, and* I *a proper ideal. Then*

$$\varepsilon(M) \subseteq I \Rightarrow I \subseteq M.$$

I is proper; choose N maximal such that $I \subseteq N$. If $N \neq M$, choose

$$a \notin N,\ b \notin M \text{ with } a\,R\,b = 0.$$

By maximality of M, $M + b\,R = R$ and thus

$$1 = m + b\,r\ ;\ m \in M\ ;\ r \in R.$$

This implies, for any $s \in R$,

$$a\,s = a\,s\,m + a\,s\,b\,r = a\,s\,m.$$

Thus $a \in \varepsilon(M) \subseteq I \subseteq N$, which is a contradiction. Finally $N = M$. ■

Corollary 14.

Let R *be a Gelfand ring,* M *a maximal ideal and* I *an ideal. Then*

$$I + M = R \Rightarrow I + \varepsilon(M) = R.$$

$I + \varepsilon(M) \supseteq \varepsilon(M)$; if $I + \varepsilon(M)$ is proper, then by lemma 13, $I + \varepsilon(M) \subseteq M$ and $I + M = M$, which is a contradiction. ■

Lemma 15.

Let R *be a Gelfand ring,* M *a maximal ideal,* I *an ideal and* $r \in R$.

$$[I : r] \subseteq M \Rightarrow I \subseteq M.$$

If $I \nsubseteq M$, $I + M = R$ and thus $I + \varepsilon(M) = R$ (corollary 14). Write

$$r = a + i\ ;\ a \in \varepsilon(M)\ ;\ i \in I.$$

Choose $\varepsilon \in M$ such that $a = a\,\varepsilon$.

$$r(1 - \varepsilon) = a(1 - \varepsilon) + i(1 - \varepsilon) = i(1 - \varepsilon) \in I.$$

This implies $1 - \varepsilon \in [I : r]$ and thus $1 - \varepsilon \in M$. But $\varepsilon \in M$ and thus $1 = (1 - \varepsilon) + \varepsilon \in M$, which is a contradiction. So $I \subseteq M$. ■

Proposition 16.

Let I *be any ideal in a Gelfand ring. Then* $\varepsilon(I)$ *is contained in* I.

Let $a \in \varepsilon(I)$ and choose $\varepsilon \in I$ such that $a = a\,\varepsilon$. Consider the ideal

$$J = [I : a] + (1 - \varepsilon)R.$$

If $J \neq R$, consider a maximal ideal $M \supseteq J$. By lemma 15, $I \subseteq M$ and thus $\varepsilon \in I \subseteq M$. But $1 - \varepsilon \in M$ and thus $1 = (1 - \varepsilon) + \varepsilon \in M$ which is a contradiction. Thus $J = R$ and

$$1 = r + (1 - \varepsilon)s \ ; \ a\,r \in I \ ; \ s \in R.$$
$$a = a\,r + a(1 - \varepsilon)s = a\,r \in I. \qquad \blacksquare$$

We are now going to produce an easier description of $\varepsilon(I)$. First of all :

Proposition 17.

Let I *be an ideal in a Gelfand ring. Then*

$$\varepsilon(I) = \cap \{\varepsilon(M) \mid M \text{ maximal};\ I \subseteq M\}.$$

Let J be the intersection on the right hand side. From $I \subseteq M$, we deduce $\varepsilon(I) \subseteq \varepsilon(M)$ and thus $\varepsilon(I) \subseteq J$. Conversely

$$\begin{aligned}
a \notin J &\Leftrightarrow \exists\, M \supseteq I;\ M \text{ maximal};\ a \notin \varepsilon(M) \\
&\Leftrightarrow \exists\, M \supseteq I;\ M \text{ maximal} \\
&\qquad \exists\, r \in R \quad \forall\, \varepsilon \in M \quad a\,r \neq a\,r\,\varepsilon \\
&\Leftrightarrow \exists\, M \supseteq I;\ M \text{ maximal} \\
&\qquad \exists\, r \in R \quad \forall\, \varepsilon \in M \quad 1 - \varepsilon \notin \text{Ann}\ a\,r \\
&\Leftrightarrow \exists\, M \supseteq I;\ M \text{ maximal} \\
&\qquad \exists\, r \in R \quad M + \text{Ann}\ a\,r \neq R \\
&\Leftrightarrow \exists\, M \supseteq I;\ M \text{ maximal} \\
&\qquad \exists\, r \in R \quad \text{Ann}\ a\,r \subseteq M \\
&\Leftrightarrow \exists\, r \in R \quad \exists\, M \text{ maximal } M \supseteq I + \text{Ann}\ a\,r \\
&\Leftrightarrow \exists\, r \in R \quad I + \text{Ann}\ a\,r \neq R.
\end{aligned}$$

Finally

$$\begin{aligned}
a \in J &\Leftrightarrow \forall\, r \in R \quad I + \text{Ann}\ a\,r = R \\
&\Leftrightarrow \forall\, r \in R \quad \exists\, \varepsilon \in I \quad 1 - \varepsilon \in \text{Ann}\ a\,r \\
&\Leftrightarrow \forall\, r \in R \quad \exists\, \varepsilon \in I \quad a\,r(1 - \varepsilon) = 0 \\
&\Leftrightarrow \forall\, r \in R \quad \exists\, \varepsilon \in I \quad a\,r = a\,r\,\varepsilon \\
&\Leftrightarrow a \in \varepsilon(I).
\end{aligned} \qquad \blacksquare$$

Proposition 18.

Let M *be a maximal ideal in a Gelfand ring*

$$\varepsilon(M) = \{a \in R \mid \exists\, \varepsilon \in M \quad a = a\,\varepsilon\}.$$

Consider $a \in R$ and $\varepsilon \in M$ with $a = a\,\varepsilon$. Then $1 - \varepsilon \in \text{Ann}\ a$ and $1 - \varepsilon \notin M$, thus M + Ann a = R and by corollary 14, $\varepsilon(M)$ + Ann a = R. Write

$$1 = \varphi + r \ ; \ \varphi \in \varepsilon(M) \ ; \ a\,r = o.$$
$$a = a\,\varphi + a\,r = a\,\varphi \in \varepsilon(M),$$

by proposition 11. ■

Proposition 20.

Let I *be an ideal in a Gelfand ring*

$$\varepsilon(I) = \{a \in R \mid \exists\ \varepsilon \in I \quad a = a\,\varepsilon\}.$$

By proposition 17

$$\begin{aligned} a \notin \varepsilon(I) &\Longleftrightarrow \exists\ M \supseteq I;\ M \text{ maximal};\ a \notin \varepsilon(M) \\ &\Longleftrightarrow \exists\ M \supseteq I;\ M \text{ maximal} \\ &\quad\ \forall\, m \in M \qquad a \neq a\,m \text{ (proposition 18)} \\ &\Longleftrightarrow \exists\ M \supseteq I;\ M \text{ maximal} \\ &\quad\ \forall\, m \in M \qquad 1 - m \notin \text{Ann } a \\ &\Longleftrightarrow \exists\ M \supseteq I;\ M \text{ maximal};\ M + \text{Ann } a \neq R. \\ &\Longleftrightarrow \exists\ M \supseteq I;\ M \text{ maximal};\ M \supseteq \text{Ann } a \\ &\Longleftrightarrow \exists\ M \text{ maximal};\ M \supseteq I + \text{Ann } a \\ &\Longleftrightarrow I + \text{Ann } a \neq R \end{aligned}$$

and finally

$$\begin{aligned} a \in \varepsilon(I) &\Longleftrightarrow I + \text{Ann } a = R \\ &\Longleftrightarrow \exists\ \varepsilon \in I \qquad 1 - \varepsilon \in \text{Ann } a \\ &\Longleftrightarrow \exists\ \varepsilon \in I \qquad a(1 - \varepsilon) = o \\ &\Longleftrightarrow \exists\ \varepsilon \in I \qquad a = a\,\varepsilon. \end{aligned}$$ ■

The next step will be to prove that $\varepsilon(I)$ is a pure ideal.

Proposition 21.

Let M *be a maximal ideal in a Gelfand ring.* $\varepsilon(M)$ *is just the pure part of* M.

$\varepsilon(M)$ is pure. Indeed, consider $a \in \varepsilon(M)$ and $\varepsilon \in M$ such that $a = a\,\varepsilon$. Hence, $1 - \varepsilon \in$ Ann a and thus M + Ann a = R. By corollary 14, $\varepsilon(M)$ + Ann a = R and thus $\varepsilon(M)$ is pure (proposition VII - 2).

Conversely, the pure part $\overset{\circ}{M}$ of M is obviously contained in $\varepsilon(M)$ (proposition 18); so $\varepsilon(M) = \overset{\circ}{M}$. ■

We now propose to generalize proposition 21 to an arbitrary ideal I. To do this, we require some left-right symmetry properties of Gelfand rings.

Lemma 22.

Let M, N *be two maximal ideals in a Gelfand ring.*

$$M = N \iff \varepsilon(M) = \varepsilon(N).$$

Immediate from lemma 13. ∎

Lemma 23.

For any maximal ideal M *in a Gelfand ring* R *and* a, b ∈ R, *we have*

$$1 - a\,b \in \varepsilon(M) \iff 1 - b\,a \in \varepsilon(M).$$

Suppose $1 - a\,b \in \varepsilon(M)$; we shall prove that $\varepsilon(M) + b\,R = R$. Indeed, if N is a maximal ideal such that

$$\varepsilon(M) + b\,R \subseteq N \neq R$$

then $\varepsilon(M) \subseteq N$ and thus by lemma 22, $M = N$ and so $b \in M$. But then $a\,b \in M$ and $1 - a\,b \in \varepsilon(M) \subseteq M$, thus $1 \in M$ which is a contradiction. Thus $\varepsilon(M) + b\,R = R$ and we can choose $r \in R$ such that $1 - b\,r \in \varepsilon(M)$. Thus

$$\begin{cases} 1 - b\,r \in \varepsilon(M) \\ 1 - a\,b \in \varepsilon(M) \end{cases} \Rightarrow \begin{cases} [b]\,.\,[r] = 1 \\ [a]\,.\,[b] = 1 \end{cases} \text{in } R/\varepsilon(M)$$

$$\Rightarrow \quad [r] = [a][b][r] = [a] \text{ in } R/\varepsilon(M)$$

$$\Rightarrow \quad r - a \in \varepsilon(M).$$

Therefore,

$$\begin{aligned} 1 - b\,a &= (1 - b\,r) + (b\,r - b\,a) \\ &= (1 - b\,r) + b(r - a) \in \varepsilon(M). \end{aligned}$$
∎

Lemma 24.

For any ideal I *in a Gelfand ring* R *and* a, b ∈ R

$$1 - a\,b \in \varepsilon(I) \iff 1 - b\,a \in \varepsilon(I).$$

By proposition 17 and lemma 23, we have

$$\begin{aligned} 1 - a\,b \in \varepsilon(I) &\iff \forall M \supseteq I,\ M \text{ maximal},\ 1 - a\,b \in \varepsilon(M) \\ &\iff \forall M \supseteq I,\ M \text{ maximal},\ 1 - b\,a \in \varepsilon(M) \\ &\iff 1 - b\,a \in \varepsilon(I). \end{aligned}$$
∎

We know that M and $\varepsilon(M)$ are 2-sided ideals for any maximal ideal M. So it makes sense to consider the equivalents of lemma 13 and corollary 14 for a left ideal I.

Lemma 25.

Let R *be a right Gelfand ring,* M *a right maximal ideal and* I *a proper left ideal. Then*

$$\varepsilon(M) \subseteq I \Rightarrow I \subseteq M.$$

Consider $a \in I$. If $\varepsilon(M) + aR = R$, choose $r \in R$ such that $1 - ar \in \varepsilon(M)$. By lemma 23, $1 - ra \in \varepsilon(M) \subseteq I$ and on the other hand, $ra \in I$ since I is left sided. Thus $1 \in I$ which is a contradiction. Thus we deduce that $\varepsilon(M) + aR$ is not the whole ring R. Consider a right maximal ideal N such that $\varepsilon(M) + aR \subseteq N$. From $\varepsilon(M) \subseteq N$, we deduce $M = N$ (lemma 22) and thus $a \in M$. ■

Corollary 26.

Let R *be a right Gelfand ring,* M *a right maximal ideal and* I *a left ideal. Then,*

$$I + M = R \Rightarrow I + \varepsilon(M) = R.$$

If $I + \varepsilon(M) \neq R$, by lemma 25, $I + \varepsilon(M) \supseteq \varepsilon(M) \Rightarrow I + \varepsilon(M) \subseteq M$ and thus $I + M = M$, which is a contradiction. ■

Lemma 27.

Let R *be a right Gelfand ring and* M *a right maximal ideal. Then*

$$(\forall a \in R)\ (\forall \varepsilon \in M)\ (a = \varepsilon a \Rightarrow a \in \varepsilon(M)).$$

M is proper, thus $1 - \varepsilon \notin M$. Consider L-Ann (a), the left annihilator of a. Then, $1 - \varepsilon \in$ L-Ann (a) thus M + L-Ann (a) = R by maximality of M (proposition 9). Hence $\varepsilon(M)$ + L-Ann (a) = R (corollary 27); choose

$$1 = r + s\ ;\ r \in \varepsilon(M)\ ;\ sa = o.$$

Now, we obtain

$$a = ra + sa = ra \in \varepsilon(M).$$ ■

Lemma 28.

Let R *be a right Gelfand ring and* I *a right ideal.*

$$(\forall a \in R)\ (\forall \varepsilon \in I)\ (a = \varepsilon a \Rightarrow a \in \varepsilon(I)).$$

By proposition 17 and lemma 27, we have

$$\begin{aligned}(a = \varepsilon a,\ \varepsilon \in I) &\Rightarrow \forall M \supseteq I,\ M \text{ maximal},\ a = \varepsilon a,\ \varepsilon \in M\\ &\Rightarrow \forall M \supseteq I,\ M \text{ maximal},\ a \in \varepsilon(M)\\ &\Rightarrow a \in \varepsilon(I).\end{aligned}$$

■

Proposition 29.

Let R *be a Gelfand ring,* I *an ideal and* M *a maximal ideal. Then,*

$$I \subseteq M \Leftrightarrow \varepsilon(I) \subseteq M.$$

From $I \subseteq M$, we deduce $\varepsilon(I) \subseteq \varepsilon(M) \subseteq M$, by proposition 12. Conversely, if $\varepsilon(I) \subseteq M$ and $I \not\subseteq M$, the maximality of M implies $I + M = R$ and thus $I + \varepsilon(M) = R$ (corollary 14). Choose

$$\begin{aligned}1 &= i + a\ ;\ i \in I\ ;\ a \in \varepsilon(M)\\ a &= a\,\varepsilon\ \ ;\ \varepsilon \in M.\end{aligned}$$

Now we obtain

$$\begin{aligned}1 - \varepsilon &= (i + a)\ (1 - \varepsilon)\\ &= i - i\,\varepsilon + a - a\,\varepsilon\\ &= i(1 - \varepsilon).\end{aligned}$$

From $i \in I$ and lemma 28, we have

$$1 - \varepsilon \in \varepsilon(I) \subseteq M$$

which is a contradiction since $\varepsilon \in M$. ■

Proposition 30.

Let R *be a Gelfand ring and* I *an ideal. Then* $\varepsilon(I)$ *is the pure part of* I.

Take $a \in \varepsilon(I)$ and consider $\varepsilon(I) + \text{Ann } a$. If $\varepsilon(I) + \text{Ann } a$ is contained in some maximal M, I is contained in M by proposition 29. Consider $\varepsilon \in I$ such that $a = a\,\varepsilon$; $\varepsilon \in I \subseteq M$ and $1 - \varepsilon \in \text{Ann } a \subseteq M$: this is a contradiction. Thus $\varepsilon(I) + \text{Ann } a = R$ and we conclude by proposition VII - 2 that $\varepsilon(I)$ is pure.

On the other hand, the pure part of I is obviously contained in $\varepsilon(I)$; thus $\varepsilon(I)$ is the pure part of I. ■

Finally, by propositions 20 and 30, the pure part of an ideal I in a Gelfand ring R is

$$\overset{\circ}{I} = \{a \in R \mid \exists\, \varepsilon \in I \quad a\,\varepsilon = a\}.$$

§ 3. CHARACTERIZATIONS OF GELFAND RINGS

We are now in a position to prove the equivalence between the notions of right Gelfand rings and left Gelfand rings : in fact, in a Gelfand ring, right maximal ideals coincide with left maximal ideals. But a Gelfand ring can also be characterized in terms of pure ideals : a ring is Gelfand if and only if the mapping which sends an ideal to its pure part is a continuous homomorphism on the lattice of ideals. From this theorem, we deduce several consequences : two of them are worth to be mentioned here. In a Gelfand ring, the "pure part" morphism and the "Jacobson radical" morphism determine a Galois connection on the lattice of ideals. Moreover, in a Gelfand ring, left pure ideals coincide with right pure ideals.

Theorem 31.

For a ring R, *the following conditions are equivalent :*

(R1) R *is a right Gelfand ring.*

(R2) *For any right maximal ideals* $M \neq N$

$$\exists\, a \notin M \quad \exists\, b \notin N \qquad a R b = o.$$

(R3) *For any right ideals* I, J

$$I + J = R \Rightarrow \overset{\circ}{I} + J = R.$$

(R4) *For any right ideals* I, J, $(I_k)_{k\in K}$

$$\overset{\circ}{\overbrace{I \cap J}} = \overset{\circ}{I} \cap \overset{\circ}{J}$$

$$\overset{\circ}{\overbrace{\underset{k\in K}{+} I_k}} = \underset{k\in K}{+} \overset{\circ}{I_k}.$$

(L1) - (L2) - (L3) - (L4) : *dual conditions of* (R1) - (R2) - (R3) - (R4).

(R1) ⇒ (R2) is just definition 4. Let us prove (R2) ⇒ (R3). If $\overset{\circ}{I} + J \neq R$, consider a maximal ideal M such that $\overset{\circ}{I} + J \subseteq M$. By proposition 29, $\overset{\circ}{I} = \varepsilon(I) \subseteq M$ and thus $I \subseteq M$. This implies $I + J \subseteq M$ which is a contradiction.

Let us prove (R3) ⇒ (R4). The condition on finite intersections is just proposition VII - 9; by the same proposition, we have the inclusion

$$\underset{k\in K}{+} \overset{\circ}{I_k} \subseteq \overset{\circ}{\overbrace{\underset{k\in K}{+} I_k}}.$$

To prove the converse inclusion, consider $a \in \overset{\circ}{\widehat{\underset{k\in K}{+} I_k}}$ and $\varepsilon \in \overset{\circ}{\widehat{\underset{k\in K}{+} I_k}}$ such that $a\,\varepsilon = a$. We can write $\varepsilon = \varepsilon_1 + \dots + \varepsilon_n$ with $\varepsilon_k \in I_k$. Since $1 - \varepsilon \in \operatorname{Ann} a$, we can write

$$I_1 + \dots + I_n + \operatorname{Ann} a = R$$

and an iterated application of (R3) yields

$$\mathring{I}_1 + \dots + \mathring{I}_n + \operatorname{Ann} a = R.$$

But then there exists $\varphi \in \mathring{I}_1 + \dots + \mathring{I}_n$ such that $1 - \varphi \in \operatorname{Ann} a$. Thus $a = a\,\varphi \in \mathring{I}_1 + \dots + \mathring{I}_n \subseteq \underset{k\in K}{+} \mathring{I}_k$.

To prove (R4) ⇒ (R1), consider two maximal ideals $M \neq N$. By (R4), we have $\mathring{M} + \mathring{N} = R$. So we can write

$$1 = m + n\ ;\ m \in \mathring{M} \text{ and } n \in \mathring{N}.$$

Then there exists $\varepsilon \in \mathring{N}$ such that $n\,\varepsilon = n$ and $1 - \varepsilon \in \operatorname{Ann} n$. Therefore

$$\mathring{N} + \operatorname{Ann} n = R \qquad (\varepsilon \in \mathring{N},\ 1 - \varepsilon \in \operatorname{Ann} n).$$

By (R4), we obtain

$$\mathring{N} + \overset{\circ}{\widehat{\operatorname{Ann} n}} = R$$

$$x + y = 1 \qquad ;\ y \in \overset{\circ}{\widehat{\operatorname{Ann} n}} \text{ and } y \notin N.$$

As $\overset{\circ}{\widehat{\operatorname{Ann} n}}$ is two-sided, we obtain

$$\forall\, r \in R \qquad r\,y \in \overset{\circ}{\widehat{\operatorname{Ann} n}} \subseteq \operatorname{Ann} n.$$

And $n\,R\,y = o$ with $n \notin M$ and $y \notin N$.

By the left-right duality, the equivalences (L1) ⟺ (L2) ⟺ (L3) ⟺ (L4) are proved. To conclude the proof, it suffices to show (R1) ⇒ (L1). This will be done if we prove that in a right Gelfand ring, any left maximal ideal is right maximal. Thus let R be a right Gelfand ring and N a maximal left ideal. We shall prove the existence of a right maximal ideal M such that $\mathring{M} \subseteq N$. By lemma 26, this will imply $N \subseteq M$ and thus $M = N$ since M is also left maximal (proposition 9).

So we must prove that a left maximal ideal N contains the pure part of some right maximal ideal M. Suppose that for every right maximal ideal M, $\mathring{M} \not\subseteq N$ and choose $a_M \in \mathring{M} \setminus N$. Consider :

$$I = + \{\text{Ann } a_M \mid M \text{ maximal}\}.$$

If $I \neq R$, fix M maximal such that $I \subseteq M$ and choose $\varepsilon \in \overset{\circ}{M}$ such that $a_M \cdot \varepsilon = a_M$. Now

$$1 - \varepsilon \in \text{Ann } a_M \subseteq I \subseteq M$$
$$\varepsilon \in \overset{\circ}{M} \subseteq M$$

which yields a contradiction. Thus $I = R$ and

$$1 = \varepsilon_1 + \ldots + \varepsilon_n \; ; \; \varepsilon_k \in \text{Ann } a_{M_k}$$

where M_k is maximal. But for each maximal ideal M, $1 \notin M$ and thus there exists some index $k(M)$ such that $\varepsilon_{k(M)} \notin M$. The maximality of M gives $M + \varepsilon_{k(M)}R = R$; so we can write

$$1 = m_M + \varepsilon_{k(M)} \cdot r_M \; ; \; m_M \in M \; ; \; r_M \in R.$$
$$a_{k(M)} = a_{k(M)} \cdot m_M + a_{k(M)} \cdot \varepsilon_{k(M)} \cdot r_M$$
$$= a_{k(M)} \cdot m_M.$$

By proposition 18, $a_{k(M)} \in \overset{\circ}{M}$. This implies that for any maximal ideal M

$$a_{M_1} R \, a_{M_2} R \ldots R \, a_{M_n} \subseteq \overset{\circ}{M}$$

and finally by proposition VII - 10 applied to (o) :

$$a_{M_1} R \, a_{M_2} R \ldots R \, a_{M_n} \subseteq \cap \overset{\circ}{M} = (o).$$

But the 2-sided part of N is a prime ideal (lemma 2, valid for an arbitrary ring and thus also for left ideals). This implies that some a_{M_k} is in the 2-sided part of N and thus in N. This contradicts the choice of a_{M_k}, so there must be some right maximal ideal M such that $\overset{\circ}{M} \subseteq N$. ∎

Corollary 32.

In a Gelfand ring, the left maximal ideals are exactly the right maximal ideals.

By (R1) ⟺ (L1) and proposition 9. ∎

Proposition 33.

In a Gelfand ring, the left pure ideals are exactly the right pure ideals.

Consider a 2-sided ideal I in the Gelfand ring R and its left pure part I^{ℓ}. By definition of a left pure ideal

$$\forall a \in I^{\ell} \qquad \exists \varepsilon \in I^{\ell} \qquad \varepsilon a = a.$$

By lemma 29, this implies that any $a \in I^{\ell}$ belongs to the right pure part I^r of I. So $I^{\ell} \subset I^r$. But the equivalence (R1) $\Leftrightarrow$ (L1) in theorem 31 implies dually $I^r \subseteq I^{\ell}$ and finally $I^r = I^{\ell}$.

Now if I is left pure, I is 2-sided and

$$I = I^{\ell} = I^r$$

which implies that I is right pure. Again by theorem 31, any right pure ideal is left pure. ■

Thus in a Gelfand ring, we can speak of pure ideals without any specification of left or right. We conclude this paragraph with a description of the relation between pure ideals and Jacobson radicals.

Proposition 34.

Let R *be a Gelfand ring and* r(R) *the lattice of right ideals of* R. *The mappings*

$$\pi : r(R) \to r(R) \quad ; \quad I \longmapsto \overset{\circ}{I}$$

$$\rho : r(R) \to r(R) \quad ; \quad I \longmapsto \text{rad } I$$

describe a Galois connection.

A dual result holds for left ideals.

By rad I, we denote the radical of I :

$$\text{rad } I = \cap \{M \mid M \text{ maximal}; M \supseteq I\}.$$

We must prove that for any two ideals I, J

$$\overset{\circ}{I} \subseteq J \Leftrightarrow I \subseteq \text{rad } J.$$

Suppose first $\overset{\circ}{I} \subseteq J$. For any maximal ideal $M \supseteq J$, the inclusion $M \supseteq \overset{\circ}{I}$ implies $M \supseteq I$ (proposition 29); thus $I \subseteq \text{rad } J$. Conversely, suppose $I \subseteq \text{rad } J$. For any maximal ideal $M \supseteq J$, $I \subseteq M$ implies $\overset{\circ}{I} \subseteq \overset{\circ}{M}$ and finally

$$\overset{\circ}{I} \subseteq \cap \{\overset{\circ}{M} \mid M \text{ maximal}; M \supseteq J\}$$

$$= \overset{\circ}{J} \qquad \text{(proposition VII - 10)}$$

$$\subseteq J.$$ ■

Corollary 35.

In a Gelfand ring, any pure ideal is the pure part of its radical.

Let I be a pure ideal. From proposition 34, we deduce :

$$\overset{\circ}{\widehat{\mathrm{rad}\ I}} \subseteq I \qquad \Longleftrightarrow \qquad \mathrm{rad}\ I \subseteq \mathrm{rad}\ I.$$

So $\overset{\circ}{\widehat{\mathrm{rad}\ I}} \subseteq I$. On the other hand, I is pure and $I \subseteq \mathrm{rad}\ I$, thus $I \subseteq \overset{\circ}{\widehat{\mathrm{rad}\ I}}$.
Finally $I = \overset{\circ}{\widehat{\mathrm{rad}\ I}}$. ■

§ 4. PURE SPECTRUM OF A GELFAND RING

In chapter 7, we defined the pure spectra of a ring : the left pure spectrum and the right pure spectrum. For a Gelfand ring, both spectra coincide. Moreover, the points of this pure spectrum are just the purely maximal ideals.
As a consequence, for Gelfand rings, the pure spectrum is homeomorphic to the usual maximal spectrum.

Proposition 36.

For a Gelfand ring, the left pure spectrum coincides with the right pure spectrum.

By proposition 33 and the definition of a pure spectrum. ■

Proposition 37.

In a Gelfand ring, any purely prime ideal is purely maximal.

Let J be a purely prime ideal in the Gelfand ring R. J is contained in some purely prime maximal ideal M (proposition VII - 28). Suppose $J \neq M$.
By theorem 31

$$M = \underset{a\in M}{+} R\,a \quad \Rightarrow \quad M = \underset{a\in M}{+} \overset{\circ}{\widehat{R\,a}}.$$

Choose $a \in M$ such that $\overset{\circ}{\widehat{R\,a}} \not\subseteq J$. By proposition VII - 2

$$M + \mathrm{Ann}\ a = R$$

and by theorem 31

$$M + \overset{\circ}{\widehat{\mathrm{Ann}\ a}} = R.$$

But by the dual of proposition VII - 16,

$$\overset{\circ}{\widehat{R\,a}} \cap \overset{\circ}{\widehat{\mathrm{Ann}\ a}} = \overset{\circ}{\widehat{R\,a}} \cap \overset{\circ}{\widehat{\mathrm{Ann}\ R\,a}}$$

$$\subseteq \overset{\circ}{\widehat{R\,a}} \cap \overset{\circ}{\widehat{\mathrm{Ann}\ \overset{\circ}{\widehat{R\,a}}}}$$

$$\subseteq \overset{\circ}{\widehat{R\,a}} \cap \mathrm{Ann}\ \overset{\circ}{\widehat{R\,a}}$$

$$= (o) \subseteq J.$$

Since J is purely prime, $\overset{\circ}{\widehat{R\,a}} \subseteq J$ or $\overset{\circ}{\widehat{\mathrm{Ann}\ a}} \subseteq J$. The choice of a implies $\overset{\circ}{\widehat{\mathrm{Ann}\ a}} \subseteq J \subseteq M$ which yields a contradiction with $M + \overset{\circ}{\widehat{\mathrm{Ann}\ a}} = R$. ■

Proposition 38.

In a Gelfand ring, the "pure part" operation induces a one-to-one correspondance between maximal ideals and purely maximal ideals.

The pure part of a maximal ideal is purely prime (proposition VII - 27) and thus purely maximal (proposition 37). Moreover this correspondance is injective (lemma 22). To prove the surjectivity, consider a purely maximal ideal J; it is contained in a maximal ideal M and thus $\overset{\circ}{J} \subseteq \overset{\circ}{M}$. But J and $\overset{\circ}{M}$ are purely maximal (first part of the proof), thus $J = \overset{\circ}{M}$. ■

Theorem 39.

The pure spectrum of a Gelfand ring is compact Hausdorff.

By proposition VII - 34, it suffices to prove that the spectrum is Hausdorff. Consider two distinct purely maximal ideals J_1, J_2 (proposition 37) which are the pure parts of two distinct maximal ideals M_1, M_2 (proposition 38). Since the ring is Gelfand

$$\exists\, a \notin M_1 \qquad \exists\, b \notin M_2 \qquad a\,R\,b = o.$$

We deduce, by theorem 31 and proposition 38,

$$M_1 + a\,R = R \Rightarrow J_1 + \overset{\circ}{\widehat{a\,R}} = R.$$

$$M_2 + R\,b = R \Rightarrow J_2 + \overset{\circ}{\widehat{R\,b}} = R.$$

In other words, $\overset{\circ}{\widehat{a\,R}} \not\subseteq J_1$ and $\overset{\circ}{\widehat{R\,b}} \not\subseteq J_2$ with

$$\overset{\circ}{\widehat{a\,R}} \cap \overset{\circ}{\widehat{R\,b}} = \overset{\circ}{\widehat{a\,R}}\ \overset{\circ}{\widehat{R\,b}} \subseteq a\,R\ R\,b = a\,R\,b = o.$$

by proposition VII - 2. By theorem 31, this means exactly :

$$J_1 \in \mathcal{O}_{\overset{\circ}{\widehat{a\,R}}} \quad ; \quad J_2 \in \mathcal{O}_{\overset{\circ}{\widehat{R\,b}}} \quad ; \quad \mathcal{O}_{\overset{\circ}{\widehat{a\,R}}} \cap \mathcal{O}_{\overset{\circ}{\widehat{R\,b}}} = \phi$$

and thus Spp(R) is Hausdorff. ■

Proposition 40.

The pure spectrum of a Gelfand ring is homeomorphic to its usual maximal spectrum.

The points of the maximal spectrum are the maximal ideals; the topology is generated by

$$\mathcal{O}_r = \{M \mid M \text{ maximal};\ r \notin M\}$$

for any $r \in R$. This is in fact a base for the topology since each maximal ideal is completely prime (proposition 8) :

$$\begin{aligned} \mathcal{O}_r \cap \mathcal{O}_s &= \{M \mid M \text{ maximal};\ r \notin M;\ s \notin M\} \\ &= \{M \mid M \text{ maximal};\ r\,s \notin M\} \\ &= \mathcal{O}_{r\,s}. \end{aligned}$$

Then proposition 38 describes a bijection between the pure spectrum and the maximal spectrum. Let us prove it is an homeomorphism. Consider $r \in R$ and M a maximal ideal. By theorem 31 :

$$\begin{aligned} M \in \mathcal{O}_r &\Leftrightarrow r \notin M \\ &\Leftrightarrow M + r\,R = R \\ &\Leftrightarrow \overset{\circ}{M} + \overset{\circ}{\widehat{r\,R}} = R \\ &\Leftrightarrow \overset{\circ}{\widehat{r\,R}} \not\subseteq \overset{\circ}{M} \\ &\Leftrightarrow \overset{\circ}{M} \in \mathcal{O}_{\overset{\circ}{\widehat{r\,R}}}. \end{aligned}$$

Thus any fundamental open subset of the maximal spectrum corresponds to an open subset of the pure spectrum.

Conversely let I be a pure ideal in R. Again by theorem 31

$$I = \underset{r \in I}{+}\ r\,R \Rightarrow I = \underset{r \in I}{+}\ \overset{\circ}{\widehat{r\,R}}.$$

Therefore by theorem VII - 32

$$\mathcal{O}_I = \bigcup_{r \in I} \mathcal{O}_{\widehat{\overset{\circ}{r} R}},$$

and the first part of the proof shows that each $\mathcal{O}_{\widehat{\overset{\circ}{r} R}}$ corresponds to a fundamental open subset in the maximal spectrum. This concludes the proof. ■

§ 5. PURE REPRESENTATION OF A GELFAND RING

In chapter 7, we described two different sheaf representations of a ring R on each of its pure spectra. For a Gelfand ring, these four representations coincide. Moreover the stalk of the representation at some point $J \in \mathrm{Spp}(R)$ is just the quotient R/J which turns out to be a local ring and the localization of R at the unique maximal ideal containing J.

Proposition 41.

For a Gelfand ring R*, the four sheaf representations* ΔR *and* ∇R *(on the right and on the left) are isomorphic.*

By proposition 38, these four representations are defined on the same topological space. Now if I is any pure ideal in R (left pure and right pure by proposition 33), the greatest left-sided ideal whose intersection with I is zero and the greatest right-sided ideal whose intersection with I is zero are 2-sided (propositions VII - 14 and 16) : thus these ideals coincide. As a consequence both sheaves ∇R, defined on the left or on the right, coincide.

To conclude the proof, it suffices to show that ΔR and ∇R, defined on the right, coincide. Let I be a pure ideal in R. The left purity of I (proposition 33) implies

$$I = \sum_{\varepsilon \in I} \{ i \in I \mid \varepsilon\, i = i \}$$

and each subset

$$I_\varepsilon = \{ i \in I \mid \varepsilon\, i = i \}$$

is obviously a right ideal. By theorem 31

$$I = \sum_{\varepsilon \in I} \overset{\circ}{I}_\varepsilon .$$

Now, if $f : I \to I$ is a right-linear endomorphism, for any $\varepsilon \in I$ and $i \in \overset{\circ}{I}_\varepsilon$:

$$f(i) = f(\varepsilon\, i) = f(\varepsilon) i.$$

Thus the restriction of f at each $\overset{\circ}{I}_\varepsilon$ is just the left multiplication by $f(\varepsilon)$.

For a pure ideal I, we are now able to define a linear mapping $\Delta R(I) \to \overset{\circ}{\nabla} R(I)$. Take some $f \in \Delta R(I) = (I, I)$; consider the pure covering $I = \underset{\varepsilon \in I}{+} \overset{\circ}{I}_\varepsilon$ and for each $\varepsilon \in I$, take $[f(\varepsilon)] \in R/C\overset{\circ}{I}_\varepsilon$. This is a compatible family; indeed choose $\varepsilon_1, \varepsilon_2 \in I$: we must prove that $f(\varepsilon_1) - f(\varepsilon_2) \in$ $\in C(\overset{\circ}{I}_{\varepsilon_1} \cap \overset{\circ}{I}_{\varepsilon_2})$. Indeed for any $i \in \overset{\circ}{I}_{\varepsilon_1} \cap \overset{\circ}{I}_{\varepsilon_2}$

$$(f(\varepsilon_1) - f(\varepsilon_2))i = f(\varepsilon_1)i - f(\varepsilon_2)i = f(\varepsilon_1\, i) - f(\varepsilon_2\, i)$$
$$= f(i) - f(i) = 0.$$

To this compatible family $[f(\varepsilon)] \in R/C\overset{\circ}{I}_\varepsilon$ corresponds a unique element in $\nabla R(I)$; this produces a mapping

$$\delta_I : \Delta R(I) \to \nabla R(I),$$

which obviously respects the additive structure of the rings. It is less obvious that δ_I respects multiplication too. Consider f, g two elements in $\Delta R(I)$. To show that $\delta_I(f \circ g) = \delta_I(f) . \delta_I(g)$, we will show that the restrictions of these elements to each $\overset{\circ}{I}_\varepsilon$, with $\varepsilon \in I$, coincide i.e. $[f(g(\varepsilon))] = [f(\varepsilon)][g(\varepsilon)]$ in $R/C\overset{\circ}{I}_\varepsilon$. Consider any element $i \in \overset{\circ}{I}_\varepsilon$. Then by right-linearity of f and g and by definition of I_ε, we obtain :

$$(f(g(\varepsilon)) - f(\varepsilon).g(\varepsilon)).i = f(g(\varepsilon).i) - f(\varepsilon).g(\varepsilon\, i)$$
$$= f(g(i)) - f(\varepsilon.g(i))$$
$$= f(g(i)) - f(g(i)) = 0.$$

The last equality follows from the fact that, since $i \in \overset{\circ}{I}_\varepsilon$ there exists $\varepsilon' \in \overset{\circ}{I}_\varepsilon$ such that $i\, \varepsilon' = i$ and thus

$$g(i) = g(i\, \varepsilon') = g(i)\varepsilon' \in \overset{\circ}{I}_\varepsilon \subseteq I_\varepsilon.$$

So $f(g(\varepsilon)) - f(\varepsilon)g(\varepsilon) \in C\,\overset{\circ}{I}_\varepsilon$.
In a similar way, we can show that $1 - \varepsilon \in C\overset{\circ}{I}_\varepsilon$. Therefore $\delta_I\,(id_I) = 1$ in $\nabla R(I)$. Thus δ_I is a ring homomorphism.

In fact, we have described a homomorphism of sheaves :

$$\delta : \Delta R \to \nabla R.$$

This means that for any pure ideals $J \subseteq I$, the following square

$$\begin{array}{ccc} \Delta R(I) & \xrightarrow{\ \delta_I\ } & \nabla R(I) \\ \big\downarrow & & \big\downarrow \\ \Delta R(J) & \xrightarrow[\ \delta_J\ ]{} & \nabla R(J) \end{array}$$

commutes, where the vertical arrows are the canonical restriction morphisms. Let f be an element in $\Delta R(I)$. Again to prove the equality

$$\delta_J(f|_J) = \delta_I(f)|_J,$$

it suffices to prove that both sides have the same restriction on each $\mathring{J}_\varepsilon$ with $\varepsilon \in J$:

$$\begin{aligned} \delta_I(f)\big|_J\big|_{\mathring{J}_\varepsilon} &= \delta_I(f)\big|_{\mathring{I}_\varepsilon}\big|_{\mathring{J}_\varepsilon} \\ &= [f(\varepsilon)]\big|_{\mathring{J}_\varepsilon} \\ &= [f(\varepsilon)] \in R/\mathcal{C}\,\mathring{J}_\varepsilon \\ &= [f|_J(\varepsilon)] \\ &= \delta_J(f|_J)\big|_{\mathring{J}_\varepsilon} . \end{aligned}$$

Thus δ is an homomorphism of sheaves.

We shall prove that each δ_I is an isomorphism and this will conclude the proof. Consider $f \in \Delta R(I) = (I, I)$ such that $\delta_I(f) = 0 \in \nabla R(\mathring{I})$. Then for any $\varepsilon \in I$, $[f(\varepsilon)] = 0$ in $R/\mathcal{C}\,\mathring{I}_\varepsilon$ or in other words, $f(\varepsilon) \in \mathcal{C}\,\mathring{I}_\varepsilon$. Thus for any $i \in \mathring{I}_\varepsilon$

$$f(i) = f(\varepsilon\ i) = f(\varepsilon).i = 0.$$

So f is zero on each $\mathring{I}_\varepsilon$ and $I = \underset{\varepsilon\in I}{+} \mathring{I}_\varepsilon$; hence f is zero on I and δ_I is injective.

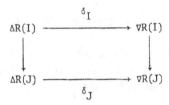

To prove that δ_I is surjective, consider some $x \in \nabla R(I)$. Then there is some pure covering $(I_k)_{k\in K}$ of I such that x can be represented by a compatible family of elements $[x_k] \in R/\mathcal{C}\,I_k$. But each pure ideal I_k itself has a pure covering given by $I_k = \underset{\varepsilon(k)\in I_k}{+} \mathring{I}_{\varepsilon(k)}$. Clearly each $[x_k]$ also gives rise to a compatible family of elements $[x_{\varepsilon(k)}] \in R/\mathcal{C}\,\mathring{I}_{\varepsilon(k)}$. Combining these facts, we see that x can be represented by a compatible family of elements $[x_\varepsilon] \in R/C\overset{\circ}{I}_\varepsilon$ for all $\varepsilon \in I$. Now, for any ε in I, define

$$f_\varepsilon : \overset{\circ}{I}_\varepsilon \to \overset{\circ}{I}_\varepsilon \; ; \; i \mapsto x_\varepsilon . i.$$

Then f_ε is a right-linear mapping defined on $\overset{\circ}{I}_\varepsilon$ and thus takes values in $\overset{\circ}{I}_\varepsilon$; hence $f_\varepsilon \in \Delta R(\overset{\circ}{I}_\varepsilon)$. Moreover, the family $(f_\varepsilon)_{\varepsilon \in I}$ is compatible. For if ε_1, ε_2 are two elements in I, then for all $i \in \overset{\circ}{I}_{\varepsilon_1} \cap \overset{\circ}{I}_{\varepsilon_2}$, we obtain :

$$\begin{aligned} f_{\varepsilon_1}(i) - f_{\varepsilon_2}(i) &= x_{\varepsilon_1}.i - x_{\varepsilon_2}.i \\ &= (x_{\varepsilon_1} - x_{\varepsilon_2})i \\ &= 0 \end{aligned}$$

since $x_{\varepsilon_1} - x_{\varepsilon_2} \in C(\overset{\circ}{I}_{\varepsilon_1} \cap \overset{\circ}{I}_{\varepsilon_2})$ by the compatibility of the family $([x_\varepsilon])_{\varepsilon \in I}$. Now the fact that the family $(f_\varepsilon)_{\varepsilon \in I}$ is compatible, implies the existence of some $f \in \Delta R(I)$ whose restriction to $\overset{\circ}{I}_\varepsilon$ is just f_ε. But $f|_{\overset{\circ}{I}_\varepsilon}(i) = f(\varepsilon)i$. So, for all $i \in \overset{\circ}{I}_\varepsilon$, we have $(x_\varepsilon - f(\varepsilon))i = 0$ and $[x_\varepsilon] = [f(\varepsilon)] \in R/C\overset{\circ}{I}_\varepsilon$. Hence $\delta_I(f) = x$ and δ_I is surjective. ∎

In order to compute the stalks of the representation $\Delta R = \nabla R$, we need the following lemma :

Lemma 42.

Let R *be a Gelfand ring and* J *a purely maximal ideal. Then the unique maximal ideal* M *containing* J *is given by*

$$\begin{aligned} M &= \{a \in R \mid \forall s \in R,\ a\,s - 1 \notin J\} \\ &= \{a \in R \mid \forall s \in R,\ s\,a - 1 \notin J\}. \end{aligned}$$

By proposition 38, we know that J is the pure part of some maximal ideal N. Now lemma 23 shows that both formulae define the same subset M of R. We will first prove that M is a 2-sided ideal.

Take a, b in M and consider a + b. For any $s \in R$,

$$(a + b)s - 1 = a\,s + b\,s - 1 = a\,s + (b\,s - 1).$$

If $(a + b)s - 1 \in J$, then $a\,s \notin J$ since $b\,s - 1 \notin J$ and similarly $b\,s \notin J$. Consider the maximal ideal N containing J. If $a\,s \in N$, then

$$a\,s \in N \quad \text{and} \quad a\,s + b\,s - 1 \in J \subseteq N$$

imply $b\,s - 1 \in N$ and thus $b\,s \notin N$. So, under the assumption $(a + b)s - 1 \in J$, we obtain

$$a\,s \notin N \quad \text{or} \quad b\,s \notin N.$$

Now take the case $a\,s \notin N$ (the case $b\,s \notin N$ can be treated similarly). Then $a \notin N$ and

$$N + a\,R = R.$$

But J is the pure part of N. So, from theorem 31, we deduce

$$J + a\,R = R.$$

In particular, there is some $r \in R$ such that $a\,r - 1 \in J$, i.e. $a \notin M$. This is a contradiction and finally $(a + b)s - 1 \notin J$ for any s. So M is stable under addition.

The relation

$$M = \{a \in R \mid \forall s \in R \quad a\,s - 1 \notin J\}$$

shows that M is stable under right multiplication and the relation

$$M = \{a \in R \mid \forall s \in R \quad s\,a - 1 \notin J\}$$

shows that M is stable under left multiplication. Finally, M is a 2-sided ideal.

M contains J because $a \in J$ and $1 \notin J$ imply $a\,s - 1 \notin J$. To conclude the proof, it suffices to show that M is a maximal ideal. Take $a \notin M$; we must prove the equality

$$M + a\,R = R.$$

But $a \notin M$ implies the existence of some $s \in R$ with $a\,s - 1 \in J$. Therefore

$$J + a\,R = R$$

and since M contains J

$$M + a\,R = R. \qquad \blacksquare$$

Theorem 43.

Let R be a Gelfand ring. The stalk of the sheaf representation $\Delta R = \nabla R$ at some point $J \in \mathrm{Spp}(R)$ is just R/J; this stalk is the localization of R at the maximal ideal containing J.

By proposition VII - 44, we know that the stalk of ∇R at J is just $R/\tilde{J}$

where

$$\mathfrak{J} = \cup \{\complement\, I \mid I \text{ pure; } I \nsubseteq J\}.$$

So we need to prove that J equals $\tilde{J}$. Since J is purely maximal (proposition 37) we know already that $\tilde{J}$ is contained in J (proposition VII - 45). Thus we must prove the inclusion $J \subseteq \mathfrak{J}$. Again we define

$$J_\varepsilon = \{j \in J \mid \varepsilon\, j = j\};$$

J_ε is a right ideal and by theorem 31 and the left purity of J

$$J = \underset{\varepsilon \in J}{+} J_\varepsilon \Rightarrow J = \underset{\varepsilon \in J}{+} \overset{\circ}{J}_\varepsilon .$$

If J is not contained in $\mathfrak{J}$, choose ε such that $\overset{\circ}{J}_\varepsilon$ is not contained in $\mathfrak{J}$. For any $j \in \overset{\circ}{J}_\varepsilon$, $(1 - \varepsilon)j = 0$ and thus $1 - \varepsilon \in \complement\, \overset{\circ}{J}_\varepsilon$. This implies that

$$J + \complement\, \overset{\circ}{J}_\varepsilon = R$$

and by theorem 31

$$J + \overset{\circ}{\complement}\, \overset{\circ}{J}_\varepsilon = R.$$

In other words, $\overset{\circ}{\complement}\, \overset{\circ}{J}_\varepsilon \nsubseteq J$ and thus by definition of $\tilde{J}$ and proposition VII - 18

$$\overset{\circ}{J}_\varepsilon \subseteq \overset{\circ}{\complement}\, \overset{\circ}{\complement}\, \overset{\circ}{J}_\varepsilon \subseteq \complement\, \overset{\circ}{\complement}\, \overset{\circ}{J}_\varepsilon \subseteq \tilde{J}$$

which contradicts the choice of ε. Thus we have already proved that the stalk at J is just $R/_J$.

Now we shall prove that $R/_J$ is a local ring obtained by inverting all the elements which are not in the maximal ideal M containing J. By proposition 38, J is the pure part of M. Consider the set N of non invertible elements in $R/_J$.

$$\begin{aligned} N &= \{[a] \mid a \in R \ \ \forall s \in R \quad [a][s] \neq 1 \quad \text{or} \quad [s][a] \neq 1\} \\ &= \{[a] \mid a \in R \ \ \forall s \in R \quad a\,s - 1 \notin J \text{ or} \quad s\,a - 1 \notin J\} \\ &= \{[a] \mid a \in R \ \ \forall s \in R \quad a\,s - 1 \notin J\} \end{aligned}$$

where the latter equality holds by lemma 24. Thus by lemma 42, the inverse image of N in R is just the maximal ideal M containing J. Therefore N is an ideal in $R/_J$ and $R/_J$ is a local ring; moreover any element which is not in M has an image which is not in N and thus is invertible in $R/_J$. Thus the elements which are not in M are exactly those which become invertible in $R/_J$.

Finally, we show that $R/_J$ is universal for the property of inverting the elements of R which are not in M. Consider a ring homomorphism $f : R \to S$ such that for any $r \notin M$, $f(r)$ is invertible in S. For any $j \in J$, choose $\varepsilon \in J \subseteq M$ such that $j\varepsilon = j$. From $\varepsilon \in M$, we deduce $1 - \varepsilon \notin M$ and thus $f(1 - \varepsilon)$ is invertible in S. But

$$f(j)f(1 - \varepsilon) = f(j(1 - \varepsilon)) = f(j - j\varepsilon) = f(j - j) = f(0) = 0$$

and since $f(1 - \varepsilon)$ is invertible, $f(j) = 0$. Finally $f(J) = 0$ and f factors uniquely through $R/_J$. ■

Theorem 43 extends a result of Bkouche ([3]) for commutative Gelfand rings. It implies in particular the following corollaries :

Corollary 44.

Let R *be a Gelfand ring and* M *a maximal ideal. The problem of localizing* R *at the prime ideal* M *has a universal solution; it is the quotient of* R *by the pure part of* M. ■

Corollary 45.

Let R *be a commutative Gelfand ring. The sheaf representation* $\Delta R = \nabla R$ *is isomorphic to that given by Bkouche in* [3].

By theorems 43 and VII - 46. ■

Bkouche's representation reduces to Pierce's representation in the case of commutative regular rings (cfr. [19]). By corollary 45, the same holds for our representation. But we shall prove a more precise result in our appendix.

§ 6. CHANGE OF BASE RING

In § 2 - 3, we proved some left-right duality conditions for Gelfand rings. They provide an effective means of transposing to Gelfand rings several results previously obtained for commutative rings. In this paragraph, we use them to establish the change of base-ring properties.

Proposition 46.

Let $f : R \to S$ *be a ring homomorphism between two Gelfand rings and* I *a pure ideal in* R. *The right sided ideal* $f(I).S$ *generated by* $f(I)$ *in* S *is pure and coincides with the left sided ideal* $S.f(I)$.

Consider the right sided ideal f(I).S. For any $i \in I$ and $s \in S$, choose $\varepsilon \in I$ such that $\varepsilon\, i = i$ (proposition 33). The equality

$$f(\varepsilon)\ f(i)s = f(i)s$$

implies that f(i)s is in the pure part of f(I).S. As a consequence f(I).S equals its pure part and thus is pure. Hence, it is also left sided and contains S.f(I). A similar argument applied to the latter finally shows that both ideals coincide. ■

This proposition 46 is the analogue of lemma VII - 60 for Gelfand rings. Let us prove the analogues of lemmas 61 and 62.

Lemma 47.

Let $f : R \to S$ *be a ring homomorphism between two Gelfand rings and* I, J *two pure ideals in* R. *In that case*

$$f(I \cap J).S = f(I).S \cap f(J).S.$$

Clearly $f(I \cap J).S \subseteq f(I).S \cap f(J).S$. Now consider

$$f(i).s = f(j).t \quad ; \quad i \in I \quad ; \quad j \in J \quad ; \quad s, t \in S.$$

Choose $\varepsilon \in I$ such that $\varepsilon\, i = i$. Then

$$\begin{aligned} f(i).s &= f(\varepsilon\, i)s \\ &= f(\varepsilon).f(i).s \\ &= f(\varepsilon).f(j).t \\ &= f(\varepsilon\, j)t \\ &\in f(I \cap J).S. \end{aligned}$$

■

Lemma 48.

Let $f : R \to S$ *be a ring homomorphism between two Gelfand rings and* M *a maximal ideal in* S. *The following equality holds*

$$\overset{\circ}{\widehat{f^{-1}(\mathring{M})}} = \overset{\circ}{\widehat{f^{-1}(M)}}$$

and this is a purely maximal ideal in R.

Clearly $\overset{\circ}{\widehat{f^{-1}(\mathring{M})}} \subseteq \overset{\circ}{\widehat{f^{-1}(M)}}$. Conversely take r in $\overset{\circ}{\widehat{f^{-1}(M)}}$. Choose ε in $\overset{\circ}{\widehat{f^{-1}(M)}}$ such that $r\,\varepsilon = r$. We deduce :

$$f(r)\ f(\varepsilon) = f(r) \quad \text{with} \quad f(r) \in M,\ f(\varepsilon) \in M.$$

By proposition 30, f(r) is in $\mathring{M}$ and thus r is in $f^{-1}(\mathring{M})$. Thus

$$\mathring{\widehat{f^{-1}(M)}} = \widehat{f^{-1}(\mathring{M})}.$$

$\mathring{M}$ is purely prime (proposition VII - 26) and the proof of lemma 62 works to show that $\mathring{\widehat{f^{-1}(\mathring{M})}}$ is purely prime (references to lemmas 60 and 61 are replaced by references to proposition 46 and lemma 47).

Therefore $\mathring{\widehat{f^{-1}(\mathring{M})}}$ is purely maximal (proposition 37). ∎

Proposition 49.

Let f : R → S *be a ring homomorphism between two Gelfand rings. Define a mapping*

$$Spp(f) : Spp(S) \to Spp(R) \; ; \; J \mapsto \mathring{\widehat{f^{-1}(J)}}.$$

This mapping is continuous.

Moreover if g : S → T *is another ring homomorphism between Gelfand rings*

$$Spp(f) \circ Spp(g) = Spp(g \circ f).$$

The proofs of propositions 63 and 64 work if the references to lemmas 55, 57, 58, 60, 61, 62 are replaced by references to proposition 46 and lemmas 47-48. ∎

§ 7. EXAMPLES OF GELFAND RINGS

We produce examples of commutative and non commutative Gelfand rings.

Example 50.

The following conditions are equivalent for a ring R.

(1) R *has a single left maximal ideal.*

(2) R *has a single right maximal ideal.*

A ring which satisfies these conditions is a Gelfand ring; in particular any local ring is a Gelfand ring.

A ring which satisfies (1) or (2) satisfies obviously conditions (L 2) or (R 2) in theorem 31; thus it is a Gelfand ring. Therefore (1) and (2) are equivalent by corollary 32. A ring is local if the non invertible elements form a 2-sided ideal; as a proper ideal never contains an invertible element, this ideal of non invertible elements is necessarily the largest proper ideal.

Example 51.

Any commutative Von Neumann regular ring is a Gelfand ring.

Let R be a commutative ring. An ideal I is regular if it is generated by its idempotent elements (cfr. example VII - 20). The ring R is Von Neumann regular if any ideal in R is regular (cfr. [19]). In particular, any ideal is pure (example VII - 20) and thus any ideal is equal to its pure part. So condition (R 3) of theorem 31 is obviously satisfied.

It should be pointed out that if any ideal of a commutative ring R is pure, R is Von Neumann regular. Indeed, for any $a \in R$, the ideal a R is pure and so there is some $r \in R$ such that $a . a r = a$. But a r is idempotent :

$$(a\ r)^2 = a^2\ r\ .\ r = a\ r\ .$$

Thus any principal ideal is regular and consequently any ideal is regular.

Example 52.

Let X *be a topological space and* $\mathbb{R}$ *the real line. The ring* $C(X, \mathbb{R})$ *of continuous functions is a commutative Gelfand ring; its pure spectrum is the Stone - Čech compactification of* X.

This is a result of Bkouche (cfr. [3]); we recall a proof below. Before this, observe that an idempotent element in $C(X, \mathbb{R})$ is a function which takes only the values 0 and 1. So if X is connected, the only two idempotent elements in $C(X, \mathbb{R})$ are 0 and 1 and Pierce's spectrum of $C(X, \mathbb{R})$ reduces to a singleton (cfr. [19]). On the other hand, the pure spectrum of $C(X, \mathbb{R})$ is the Stone - Čech compactification of X (see below), which retains a lot of relevant information on X and thus on $C(X, \mathbb{R})$. Example VII - 21 describes some pure ideals in $C(X, \mathbb{R})$.

We will now outline a proof of the fact that $C(X, \mathbb{R})$ is a Gelfand ring; we refer largely to [8]. First of all, there is always a completely regular space Y such that $C(X, \mathbb{R})$ is isomorphic to $C(Y, \mathbb{R})$ ([8] - 3 - 8). Now for any $f \in C(Y, \mathbb{R})$ define

$$Z_f = \{y \in Y \mid f(y) = 0\}.$$

The subsets Z_f form a lattice Z :

$$Z_f \cap Z_g = Z_{|f| + |g|}$$

$$Z_f \cup Z_g = Z_{fg}.$$

The maximal ideals of $C(Y, \mathbb{R})$ are in one-to-one correspondance with the Z-ultrafilters ([8] - 2 - 5); a maximal ideal $M \subseteq C(Y, \mathbb{R})$ and the corresponding Z-ultrafilter $\mathcal{U}$ are related by the formulae

$$M = \{f \in C(X, \mathbb{R}) \mid Z_f \in \mathcal{U}\}$$

$$\mathcal{U} = \{Z_f \mid f \in M\}.$$

The Stone - Čech compactification of Y (which is also that of X) is the set of all Z-ultrafilters provided with the topology admitting the subsets C_f as a base for the closed subsets ([8] - 6 - 5).

$$C_f = \{\mathcal{U} \mid \mathcal{U}\ Z\text{-ultrafilter};\ Z_f \in \mathcal{U}\}.$$

In terms of maximal ideals and open subsets, the Stone - Čech compactification of X and Y is the set of maximal ideals in $C(Y, \mathbb{R})$ provided with the topology admitting the subsets $\mathcal{O}_f$ as a base

$$\mathcal{O}_f = \{M \mid M \text{ maximal ideal in } C(Y, \mathbb{R});\ f \notin M\}.$$

But this is exactly the description of the maximal spectrum of $C(Y, \mathbb{R})$. This space is Hausdorff ([8] - 6 - 5), and thus if M, N are distinct maximal ideals, there exists $f, g \in C(Y, \mathbb{R})$ such that

$$M \in \mathcal{O}_f \quad ; \quad N \in \mathcal{O}_g \quad ; \quad \mathcal{O}_f \cap \mathcal{O}_g \neq \phi.$$

Because any maximal ideal is prime

$$\mathcal{O}_f \cap \mathcal{O}_g = \mathcal{O}_{fg}.$$

On the other hand (cfr. [8]).

$$\begin{aligned} \mathcal{O}_{fg} = \phi &\iff \forall M \text{ maximal ideal } fg \in M \\ &\iff fg \in \mathrm{rad}(0) \\ &\iff \exists n \quad (fg)^n = 0 \\ &\iff fg = 0. \end{aligned}$$

Finally the Hausdorff condition on the maximal spectrum can be written under the form

$$\exists f \notin M \qquad \exists g \notin N \qquad fg = 0$$

which is just the definition of a Gelfand ring.

Example 53.

Let R *be a commutative Gelfand ring.*

Let $\mathcal{R}$ *be the ring of* n × n *triangular matrices on* R *such that all the elements on the diagonal are equal.*

$\mathcal{R}$ *is a Gelfand ring with the same pure spectrum as* R.

Thus the elements in $\mathcal{R}$ have the form

$$\begin{pmatrix} a & & & \\ & a & \ddots & a_{ij} \\ & & \ddots & \\ 0 & & & a \end{pmatrix}.$$

The addition or multiplication of matrices in $\mathcal{R}$ is performed pointwise on the diagonal elements. Therefore if $\mathcal{I}$ is an ideal in $\mathcal{R}$, the subset

$$\alpha(\mathcal{I}) = \{a \in R \mid a \text{ appears on the diagonal of some matrix of } \mathcal{I}\}$$

is an ideal in R. This ideal is necessarily 2-sided since R is commutative. On the other hand, if I is an ideal in R, the subset

$$\beta(I) = \{A \in \mathcal{R} \mid \text{the diagonal element of } A \text{ is in } I\}$$

is a 2-sided ideal in $\mathcal{R}$. Moreover, if I is maximal in R, $\beta(I)$ is maximal in $\mathcal{R}$: indeed the only way to add a matrix to $\beta(I)$ is to introduce new diagonal elements and I, the ideal of diagonal elements, is already maximal. Now take M, N two distinct maximal ideals in R; because R is Gelfand :

$$\exists\, a \notin M \qquad \exists\, b \notin N \; : \; a\, b = 0.$$

Now consider

$$\bar{a} = \begin{pmatrix} a & & & 0 \\ & a & & \\ & & \ddots & \\ 0 & & & a \end{pmatrix} \quad ; \quad \bar{b} = \begin{pmatrix} b & & & 0 \\ & \ddots & & \\ & & \ddots & \\ 0 & & & b \end{pmatrix}.$$

For any matrix A in $\mathcal{R}$

$$\bar{a} \notin \beta(M) \qquad \bar{b} \notin \beta(N) \qquad \bar{a}\, A\, \bar{b} = 0.$$

Thus to prove that $\mathcal{R}$ is a Gelfand ring, it suffices to prove that any maximal ideal in $\mathcal{R}$ has the form $\beta(M)$ with M maximal in R. This will be done if we prove that any proper ideal $\mathcal{I}$ of $\mathcal{R}$ is contained in some $\beta(M)$ with M maximal in R.

Take a proper right ideal I in $\mathcal{R}$. Obviously we have $I \subseteq \beta(\alpha(I))$; if $\alpha(I)$ is proper in R, choose M maximal in R such that $\alpha(I) \subseteq M$. This implies $I \subseteq \beta(M)$. So the problem reduces to prove that $\alpha(I)$ is a proper ideal. If it is not, $1 \in \alpha(I)$ and thus there is some matrix $A \in I$ with 1 as a diagonal element. Now denote by I_{ij} the matrix whose (i,j)-entry is 1, while all the other entries are 0. For any $j > 1$, we have

$$I_{1j} = A \,.\, I_{1j} \in I.$$

We deduce that

$$B = A - \sum_{j=2}^{n} I_{1j} \,.\, a_{1j} \in I$$

and $b_{1j} = 0$ for $j \neq 1$. Therefore, for any $j > 2$

$$I_{2j} = B \,.\, I_{2j} \in I.$$

Iterating the process, we find that any matrix I_{ij} with $j > i$ is in I. Therefore the identity matrix is also in I :

$$1 = A - \sum_{i<j} I_{ij} \,.\, a_{ij}$$

thus I is not proper. Hence, $\alpha(I)$ must be proper, which shows that $\mathcal{R}$ is a Gelfand ring.

The pure spectrum of a Gelfand ring is just its maximal spectrum (proposition 40), thus β describes a bijection between Spp(R) and Spp($\mathcal{R}$). Let us prove it is an homeomorphism. For any $a \in R$ and any maximal ideal M in R

$$a \notin M \iff \bar{a} \notin \beta(M).$$

Thus any fundamental open subset 0_a of Spp(R) corresponds to an open subset in Spp($\mathcal{R}$). Conversely take $A \in \mathcal{R}$ with diagonal element a; for any maximal ideal M in R

$$A \notin \beta(M) \iff a \notin M.$$

Thus any fundamental open subset in Spp($\mathcal{R}$) corresponds to an open subset in R. Finally we have shown that Spp(R) and Spp($\mathcal{R}$) are homeomorphic.

We will conclude this example with a remark. In a Gelfand ring, all maximal ideals are 2-sided (proposition 7) but an arbitrary ideal need not be 2-sided. Consider in $\mathcal{R}$ the matrices with some fixed column zero (and thus diagonal zero) : this is obviously a left ideal which is not right sided. Conversely the matrices with some fixed row zero (and thus diagonal zero) form a right ideal which is not left sided.

Example 54.

Let R be a commutative Gelfand ring such that each element of the form

$$1 + (a_1)^2 + \dots + (a_n)^2$$

is invertible; the rings $C(X, \mathbb{R})$ of continuous functions satisfy this condition.

In that case, the ring of complexes of R and the ring of quaternions of R are Gelfand rings with the same pure spectrum as R.

A continuous function $f \in C(X, \mathbb{R})$ is invertible if and only if for any $x \in X$, $f(x) \neq 0$. Thus $1 + (f_1)^2 + \dots + (f_n)^2$ is always invertible.

If R is a commutative ring, the ring of complexes of R is the ring $\mathbb{C}(R)$ of matrices of the form

$$\begin{pmatrix} a & b \\ -b & a \end{pmatrix}$$

with a, b in R. This ring is obviously commutative and the transposition is a ring involution on $\mathbb{C}(R)$. The ring $\mathbb{H}(R)$ of quaternions on R is the ring of matrices of the form

$$\begin{pmatrix} A & B \\ -B^t & A^t \end{pmatrix}$$

where A, B are in $\mathbb{C}(R)$; this ring is generally not commutative. Finally let us point out that the condition on R :

"$1 + (a_1)^2 + \dots + (a_n)^2$ is invertible",

implies the following condition on $\mathbb{C}(R)$:

"$1 + A_1 \overline{A_1} + \dots + A_n \overline{A_n}$ is invertible".

Indeed

$$A_k \overline{A_k} = \begin{pmatrix} a_k & b_k \\ -b_k & a_k \end{pmatrix} \begin{pmatrix} a_k & -b_k \\ b_k & a_k \end{pmatrix} = \begin{pmatrix} a_k^2 + b_k^2 & 0 \\ 0 & a_k^2 + b_k^2 \end{pmatrix}$$

and therefore

$$1 + \sum_k A_k \overline{A_k} = \begin{pmatrix} 1 + \sum_k a_k^2 + \sum_k b_k^2 & 0 \\ 0 & 1 + \sum_k a_k^2 + \sum_k b_k^2 \end{pmatrix}$$

which is invertible since the element on the diagonal is invertible.

So the two steps in the construction of $\mathbb{H}(R)$ fit into the common following context :
"Let R be a commutative ring and $(\bar{\ })$: $R \to R$ a ring involution such that any element of the form $1 + a_1 \bar{a}_1 + \ldots + a_n \bar{a}_n$ is invertible in R. From this we construct the ring $\mathcal{R}$ of matrices of the form

$$\begin{pmatrix} a & b \\ -\bar{b} & \bar{a} \end{pmatrix} \quad \text{with a, b in R".}$$

Choosing the identity as involution on R, we obtain $\mathbb{C}(R)$; choosing transposition as involution on $\mathbb{C}(R)$, we obtain $\mathbb{H}(R)$. We will show later on that both involutions respect maximal ideals, i.e. for every maximal ideal M and $m \in M$, we have $\bar{m} \in M$. To prove the statements of our example, it then suffices to show that in the situation described above, the following implication holds :
"R is a Gelfand ring and the involution respects maximal ideals $\Rightarrow$ $\mathcal{R}$ is a Gelfand ring with same pure spectrum as R".
Let us prove the implication.

Let I be a right ideal in $\mathcal{R}$. Then define :

$$\alpha(I) = \{a \in R \mid \exists\, b \in R \begin{pmatrix} a & b \\ -\bar{b} & \bar{a} \end{pmatrix} \in I\};$$

$\alpha(I)$ is obviously an ideal in R. Moreover the relation

$$\begin{pmatrix} b & -a \\ \bar{a} & \bar{b} \end{pmatrix} = \begin{pmatrix} a & b \\ -\bar{b} & \bar{a} \end{pmatrix} \begin{pmatrix} 0 & -1 \\ 1 & 0 \end{pmatrix} \in I$$

shows that in fact

$$\begin{pmatrix} a & b \\ -\bar{b} & \bar{a} \end{pmatrix} \in I \quad \Rightarrow \quad a \in \alpha(I) \text{ and } b \in \alpha(I).$$

Conversely if I is an ideal in R, define

$$\beta(I) = \{\begin{pmatrix} a & b \\ -\bar{b} & \bar{a} \end{pmatrix} \in \mathcal{R} \mid a \in I \; ; \; b \in I\} \; ;$$

$\beta(I)$ is obviously a right ideal in $\mathcal{R}$ and if I is proper so is $\beta(I)$.

If M is a maximal ideal in R, $\beta(M)$ is necessarily a maximal ideal in $\mathcal{R}$. Indeed, consider a right ideal $\mathcal{I} \supseteq \beta(M)$ and $A \in \mathcal{I} \setminus \beta(M)$

$$A = \begin{pmatrix} a & b \\ -\bar{b} & \bar{a} \end{pmatrix} \; ; \quad a \notin M \;\text{ or }\; b \notin M.$$

According to the relation pointed out before, it suffices to consider the case $a \notin M$. Since M is maximal, we then obtain

$$M + aR = R,$$

and hence

$$m + ar = 1 \; ; \; m \in M \; ; \; r \in R.$$

$$\begin{pmatrix} 1 & b\bar{r} \\ -\bar{b}r & 1 \end{pmatrix} = \begin{pmatrix} m & 0 \\ 0 & \bar{m} \end{pmatrix} + \begin{pmatrix} a & b \\ -\bar{b} & \bar{a} \end{pmatrix} \begin{pmatrix} r & 0 \\ 0 & \bar{r} \end{pmatrix} \in \mathcal{I}$$

$$\begin{pmatrix} 1 & 0 \\ 0 & 1 \end{pmatrix} = \begin{pmatrix} 1 & b\bar{r} \\ -\bar{b}r & 1 \end{pmatrix} \begin{pmatrix} \frac{1}{1+b\bar{b}r\bar{r}} & \frac{-b\bar{r}}{1+b\bar{b}r\bar{r}} \\ \frac{\bar{b}r}{1+b\bar{b}r\bar{r}} & \frac{1}{1+b\bar{b}r\bar{r}} \end{pmatrix} \in \mathcal{I}$$

and thus $\mathcal{I} = \mathcal{R}$. So $\beta(M)$ is maximal.

If $\mathcal{I}$ is a proper ideal in $\mathcal{R}$, $\alpha(\mathcal{I})$ is necessarily a proper ideal. Indeed if $\alpha(\mathcal{I}) = R$, there is some matrix

$$\begin{pmatrix} 1 & b \\ -\bar{b} & 1 \end{pmatrix} \in \mathcal{I}$$

and thus

$$\begin{pmatrix} 1 & 0 \\ 0 & 1 \end{pmatrix} = \begin{pmatrix} 1 & b \\ -\bar{b} & 1 \end{pmatrix} \begin{pmatrix} \frac{1}{1+b\bar{b}} & \frac{-b}{1+b\bar{b}} \\ \frac{\bar{b}}{1+b\bar{b}} & \frac{1}{1+b\bar{b}} \end{pmatrix} \in I.$$

This contradicts the fact that I is proper. So $\alpha(I)$ is a proper ideal and is contained in some maximal ideal M. But then I is contained in $\beta(M)$. This shows that the maximal ideals of $\mathcal{R}$ are exactly the $\beta(M)$ with M maximal in R.

Let us prove that $\mathcal{R}$ is a Gelfand ring. Take M and N two distinct maximal ideals in R. Because R is Gelfand :

$$\exists\, a \notin M \ \exists\, b \notin N \qquad ab = 0.$$

Now $a \notin M$, $a\bar{a} \notin M$, thus

$$\begin{pmatrix} a\bar{a} & 0 \\ 0 & \bar{a}a \end{pmatrix} \notin \beta(M) \qquad \begin{pmatrix} b & 0 \\ 0 & \bar{b} \end{pmatrix} \notin \beta(N)$$

and for every $A \in \mathcal{R}$, we have

$$\begin{pmatrix} a\bar{a} & 0 \\ 0 & \bar{a}a \end{pmatrix} A \begin{pmatrix} b & 0 \\ 0 & \bar{b} \end{pmatrix} = 0.$$

Thus $\mathcal{R}$ is a Gelfand ring.

Now let us compute the pure spectrum of $\mathcal{R}$. As R and $\mathcal{R}$ are Gelfand rings, their pure spectrum is their maximal spectrum (proposition 40). The mapping β describes a one-to-one correspondance between the maximal ideals in R and the maximal ideals in $\mathcal{R}$. Let us prove that it is an homeomorphism. For any $a \in R$ and any maximal ideal M in R

$$a \notin M \iff \begin{pmatrix} a & 0 \\ 0 & \bar{a} \end{pmatrix} \notin \beta(M).$$

Thus any fundamental open subset in Spp(R) corresponds to an open subset in Spp($\mathcal{R}$). Conversely if a, b are in R and M is maximal in R,

$$A = \begin{pmatrix} a & b \\ -\bar{b} & \bar{a} \end{pmatrix} \notin \beta(M) \iff a \notin M \quad \text{or} \quad b \notin M.$$

Thus the fundamental open subset of Spp($\mathcal{R}$) generated by the element $A \in \mathcal{R}$

corresponds to the union of the open subsets of Spp(R) generated by a and b : this is an open subset in Spp(R). Finally β describes an homeomorphism between Spp(R) and Spp(R).

All that remains to be shown now is that the two particular involutions giving rise to $\mathbb{C}(R)$ and $\mathbb{H}(R)$ respect maximal ideals. Clearly the identity does so and $\mathbb{C}(R)$ is a Gelfand ring. Now any maximal ideal in $\mathbb{C}(R)$ is of the form $\beta(M)$ with M maximal in R (observe that the proof of this fact is done entirely without the additional assumption on the involution). Let $\begin{pmatrix} a & b \\ -b & a \end{pmatrix}$ be an element of $\beta(M)$ in $\mathbb{C}(R)$; thus $a \in M$ and $b \in M$. But then $-b \in M$ and $\begin{pmatrix} a & -b \\ b & a \end{pmatrix} \in \beta(M)$.

This example 54 should be compared with example 2 - 20 in [18]. Orzech and Small prove that under the weaker condition that 2 is invertible in the commutative ring R, the lattice of ideals of R is isomorphic to the lattice of 2-sided ideals of the quaternion ring $\mathbb{H}(R)$ on R.

APPENDIX : NOTE ON PIERCE'S REPRESENTATION THEOREM

Using central idempotents, Pierce proposes a sheaf representation theorem for any ring R (cfr. [19]). In this appendix, we show that the functorial description of this representation (= mapping on the open subsets of the spectrum) is just the sheaf of linear endomorphisms of pure ideals. In particular when pure ideals are exactly the regular ideals (for example in the case of von Neumann regular rings or in the case of principal rings) Pierce's representation coincides with our representation ΔR.

Definition 1.

Let R be a ring. An ideal I is regular if and only if

$$\forall i \in I \qquad \exists \varepsilon \in I \cap \textit{center } R \qquad \varepsilon = \varepsilon^2 \qquad i = \varepsilon i.$$

Proposition 2.

The definition of a regular ideal is left-right symmetric; in particular any regular ideal is 2-sided.

With the notations of definition 1, $i = \varepsilon i = i \varepsilon$ because ε is central. In particular, for any $r \in R$ and $i \in I$:

$$r i = r i \varepsilon = \varepsilon r i \in I.$$

■

Proposition 3.

Any regular ideal is pure. The converse holds for von Neumann regular rings and commutative principal rings.

A regular ideal is pure by definition. A ring is said to be regular if any ideal is regular and thus pure : hence, in that case all ideals are pure and regular. If the ring R is commutative and principal, choose a pure ideal aR ; by purity of aR

$$\exists r \in R \qquad a = a \,.\, ar$$

and ar is necessarily an idempotent element :

$$(ar)^2 = a^2 r \,.\, r = ar.$$

Thus aR is regular.

■

Proposition 4.

Let R be a ring. The set B of central idempotents in R is a boolean algebra for the binary operations $\wedge$ *and* $\vee$ *given by*

$$\varepsilon \wedge \varepsilon' = \varepsilon\varepsilon'$$
$$\varepsilon \vee \varepsilon' = \varepsilon + \varepsilon' - \varepsilon\varepsilon'.$$

The points of the corresponding Stone space X *are the maximal regular ideals of* R *and the open subsets are the* $\mathcal{O}_I$ *given by*

$$\mathcal{O}_I = \{M \mid M \text{ maximal regular ideal in } R;\ I \not\subseteq M\}$$

for any regular ideal I. X *is called the Pierce's spectrum of* R.

B is a boolean algebra and the points of the corresponding Stone space are the maximal ideals of B; a base of its topology is given by

$$\mathcal{O}_\varepsilon = \{N \mid N \text{ maximal ideal in } B;\ \varepsilon \notin N\}$$

for any central idempotent ε (cfr. [19]). We recall that $J \subseteq B$ is an ideal in B if

$$\varepsilon \in J \text{ and } \varepsilon' \in J \Rightarrow \varepsilon \vee \varepsilon' \in J$$
$$\varepsilon \in J \text{ and } \varepsilon' \in B \Rightarrow \varepsilon\varepsilon' \in J.$$

There is an obvious isomorphism between the lattice of regular ideals I in R and the lattice of ideals J in B : to I we associate the set J of its central idempotents; to J we associate the ideal it generates in R. So the space X can equivalently be defined as the set of maximal regular ideals in R. The fundamental open subsets are now

$$\mathcal{O}_\varepsilon = \{M \mid M \text{ maximal regular ideal};\ \varepsilon \notin M\}$$

for any central idempotent ε. If $(\varepsilon_k)_{k\in K}$ is a family of central idempotents and I is the regular ideal generated by this family

$$\begin{aligned}\bigcup_{k\in K} \mathcal{O}_{\varepsilon_k} &= \{M \mid M \in X\ ;\ \exists\, k \in K \quad \varepsilon_k \notin M\}\\ &= \{M \mid M \in X\ ;\ I \not\subseteq M\}\\ &= \mathcal{O}_I.\end{aligned}$$

Thus any open subset in X has the form $\mathcal{O}_I$. Conversely if I is any regular ideal and $(\varepsilon_k)_{k\in K}$ is the family of central idempotents in I, the same equalities show that $\mathcal{O}_I$ is open in X. Finally the subsets $\mathcal{O}_I$ are exactly the open subsets of X. ∎

Theorem 5 (Pierce's representation theorem - cfr. [19]).

Let R *be a ring and* X *its Pierce's spectrum. Consider the set* $\mathcal{R}$*, disjoint union of the quotient rings* $R/_M$ *for any maximal regular ideal* M*. Provide* $\mathcal{R}$ *with the final topology for all the mappings*

$$X \to \mathcal{R} \ ; \ M \longmapsto [r] \in R/_M$$

for any $r \in R$*.*

The canonical projection $p : \mathcal{R} \to X$ *is a local homeomorphism and presents* $\mathcal{R}$ *as a sheaf of rings on* X*. The ring* R *is isomorphic to the ring of continuous sections of* p*.* ∎

Theorem 6.

Let R *be a ring and* X *its Pierce's spectrum. The mapping defined by*

$$\mathcal{O}_I \longmapsto (I, I)$$

for any regular ideal I*, is a sheaf of rings on* X*. The corresponding "espace étalé" is just Pierce's spectrum as defined in theorem 5.*

The mapping

$$\mathcal{O}_I \mapsto (I, I)$$

is a sheaf of rings; indeed the proof of our theorem VII - 38 is valid if we replace the word "pure" by the word "regular". In particular, if ε is some central idempotent, a linear mapping

$$f : \varepsilon R \to \varepsilon R$$

is just the left multiplication by $f(\varepsilon)$

$$\begin{aligned} f(\varepsilon r) &= f(\varepsilon\varepsilon r) \\ &= f(\varepsilon) \,.\, \varepsilon r. \end{aligned}$$

Therefore we have an isomorphism of rings

$$(\varepsilon R, \varepsilon R) \cong \varepsilon R.$$

Now if ε' is another central idempotent such that $\varepsilon' R \subseteq \varepsilon R$, there exists $r \in R$ such that $\varepsilon' = \varepsilon r$ and the restriction of $f \in (\varepsilon R, \varepsilon R)$ to $\varepsilon' R$ is

$$\begin{aligned} f(\varepsilon' s) &= f(\varepsilon'\varepsilon' s) \\ &= f(\varepsilon\varepsilon r)\varepsilon' s \\ &= \varepsilon' f(\varepsilon)\varepsilon' s. \end{aligned}$$

In other words, via the isomorphisms

$$(\varepsilon R, \varepsilon R) \cong \varepsilon R$$

the restriction mappings are just

$$\varepsilon R \to \varepsilon' R \quad ; \quad \varepsilon r \mapsto \varepsilon' \varepsilon r.$$

Now let us consider the sheaf of rings defined in theorem 5; we shall compute the continuous sections of p on some fundamental open subset $\mathcal{O}_\varepsilon$ of X for a central idempotent $\varepsilon \in R$. Any mapping

$$\mathcal{O}_\varepsilon \hookrightarrow X \to R \quad ; \quad M \mapsto [r] \in R/M$$

for any $r \in R$ is continuous. Conversely any continuous section of p :

$$\sigma : \mathcal{O}_\varepsilon \to \mathcal{R}$$

is of this form. Indeed ε has a complement $1-\varepsilon$ in B and thus the complement of $\mathcal{O}_\varepsilon$ in X is just $\mathcal{O}_{1-\varepsilon}$ (cfr. [19]); if we define σ to be 0 on $\mathcal{O}_{1-\varepsilon}$, we obtain a continuous extension of σ

$$\overline{\sigma} : X \to \mathcal{R}.$$

By theorem 5, $\overline{\sigma}$ has the form

$$M \mapsto [r] \in R/M$$

for some $r \in R$. Finally any continuous section on $\mathcal{O}_\varepsilon$ has the form

$$\mathcal{O}_\varepsilon \to \mathcal{R} \; ; \; M \mapsto [r] \in R/M.$$

But different elements $r, s \in R$ could induce the same continuous section. This is the case if

$$\forall M \in \mathcal{O}_\varepsilon \quad , \quad r-s \in M$$

or in other words

$$\forall M \in X \quad (\varepsilon \notin M \Rightarrow r-s \in M).$$

But we know (cfr. [11]) that

$$\forall \varepsilon \in B \quad (\varepsilon \notin M \Leftrightarrow 1-\varepsilon \in M).$$

Thus finally $r, s \in R$ induce the same continuous section on $\mathcal{O}_\varepsilon$ if and only if

$$\forall M \in X \quad (1-\varepsilon \in M \Rightarrow r-s \in M),$$

or in other words

$$r-s \in \cap \{M \mid M \in X; \; 1-\varepsilon \in M\}.$$

But in a boolean algebra, any ideal is the intersection of the maximal ideals containing it (cfr. [11]); thus the principal ideal $(1-\varepsilon)R$ in R is the intersection of the maximal regular ideals containing it. Finally two elements $r, s \in R$ generate the same continuous section of p on $\mathcal{O}_\varepsilon$ if and only if

$$r-s \in (1-\varepsilon)R.$$

Therefore the ring of continuous sections of p on $\mathcal{O}_\varepsilon$ is

$$R/(1-\varepsilon)R \cong \varepsilon R.$$

Finally if ε' is another central idempotent and $\varepsilon' = \varepsilon r$ for some $r \in R$, we have obviously $\mathcal{O}_{\varepsilon'} \subseteq \mathcal{O}_\varepsilon$. Let us consider a continuous section of p on $\mathcal{O}_\varepsilon$; we have seen it has the form

$$\mathcal{O}_\varepsilon \to R \; ; \; M \mapsto [\varepsilon r] \in R/M$$

for some $r \in R$. Its restriction on $\mathcal{O}_{\varepsilon'}$ is just

$$\mathcal{O}_{\varepsilon'} \to R \; ; \; M \mapsto [\varepsilon' \varepsilon r] \in R/M$$

simply because

$$\varepsilon r - \varepsilon'\varepsilon r = (1-\varepsilon')\varepsilon r \in (1-\varepsilon')R.$$

So, up to the isomorphism described above the restriction mapping is just

$$\varepsilon R \to \varepsilon' R \; ; \; \varepsilon r \mapsto \varepsilon'\varepsilon r.$$

We are now able to conclude. The sheaves given by theorems 5 and 6 coincide on a base of the topology of X : thus they are isomorphic. ■

Finally we have shown that Pierce's representation of rings is just the analogue with regular ideals of our representation ΔR using pure ideals. Let us conclude by pointing out that Pierce's representation is also the analogue of our representation ∇R.

Corollary 7.

Let R be a ring and X its Pierce's spectrum. Pierce's sheaf is the sheaf associated to the separated presheaf

$$\mathcal{O}_I \mapsto R/C\, I$$

for any regular ideal I in R.

Replacing the word "pure" by "regular", the proofs of our propositions VII - 40 - 41 show that

$$\mathcal{O}_I \mapsto R/C\, I$$

is a separated presheaf on X.

If ε is some central idempotent in R,

$$\varepsilon R \oplus (1-\varepsilon)R = R$$

and thus $\complement\, \varepsilon R = (1-\varepsilon)R$. In other words

$$R/\complement\, \varepsilon R \cong R/(1-\varepsilon)R \cong \varepsilon R$$

and the restrictions are induced by the quotient maps. Going back to the proof of theorem 5, we conclude that this separated presheaf coincides with Pierce's sheaf on a base of the topology of X. Thus the sheaf associated to this separated presheaf is just Pierce's sheaf. ■

Corollary 8.

Let R *be a ring in which any pure ideal is regular (cfr. proposition* 3*). In that case, Pierce's spectrum of* R *coincides with the pure spectrum of* R *and Pierce's representation is isomorphic to our representations* ΔR *and* ∇R. ■

INDEX

NOTATIONS

BIBLIOGRAPHY

[1] ARTIN - GROTHENDIECK, Cohomologie étale des schémas, Sém. de géom. alg. 4, 1963/64, IHES Paris.

[2] BARR, Exact categories, Lecture notes in math. 236, Springer (1970), 1-120.

[3] BKOUCHE, Pureté, mollesse et paracompacité, C.R. Acad. Sci. Paris 270, (1970), A 1653-A 1655.

[4] DIVINSKI, Rings and radicals, Mathematical expositions 14, Univ. of Toronto Press.

[5] FAITH, Algebra II : Ring theory, Grund. der Math. Wiss. 191, Springer (1976).

[6] GABRIEL, Des catégories abéliennes, Bull. Soc. Math. de France 90, (1962), 323-448.

[7] GABRIEL - ULMER, Lokal präsentierbare Kategorien, Lect. Notes in Math. 221, Springer (1971).

[8] GILLMAN - JERISON, Rings of continuous functions, Grad. Texts in Math. 43, Springer (1976).

[9] GODEMENT, Théorie des faisceaux, Hermann (1958).

[10] GOLAN, Localization of non commutative rings, Marcel Dekker (1975).

[11] GRÄTZER, General lattice theory, Birkhaüser (1978).

[12] JOHNSTONE, Topos theory, Academic Press (1977).

[13] LAMBECK, On the representation of modules by sheaves of factor modules, Can. Math. Bull. 14-3 (1971), 359-368.

[14] LAWVERE, Functorial semantics of algebraic theories, Proc. Nat. Ac. Sci. 50 (1963), 869-872.

[15] LINTON, Autonomous categories and duality of functors, J. of alg. 2 (1965), 315-341.

[16] MULVEY, A generalization of Swan's theorem, Math. Zeit. 151 (1976), 57-70.

[17] MULVEY, Representations of rings and modules, Lect. Notes in Math. 753, Springer (1979).

[18] ORZECH - SMALL, The Brauer group of commutative rings, Marcel Dekker (1975).

[19] PIERCE, Modules over commutative regular rings, Mem. Am. Math. Soc. 70 (1967).

[20] POPESCU, Abelian categories with applications to rings and modules, Academic Press (1973).

[21] SCHUBERT, Categories, Springer (1972).

[22] SIMMONS, The frame of localizations of a ring, Dep. of Math., Univ. of Aberdeen (1981).

ADDITIONAL REFERENCE

JOHNSTONE-WRAITH, Algebraic Theories in Toposes, in Indexed Categories and their applications, Lect. Notes in Math. 661, Springer (1978).